191 Current Topics in Microbiology and Immunology

Editors

A. Capron, Lille · R.W. Compans, Atlanta/Georgia
M. Cooper, Birmingham/Alabama · H. Koprowski,
Philadelphia · I. McConnell, Edinburgh · F. Melchers, Basel
M. Oldstone, La Jolla/California · S. Olsnes, Oslo
M. Potter, Bethesda/Maryland · H. Saedler, Cologne
P.K. Vogt, La Jolla/California · H. Wagner, Munich
I. Wilson, La Jolla/California

Measles Virus

Edited by Volker ter Meulen
and Martin A. Billeter

With 23 Figures and 3 Tables

Springer-Verlag
Berlin Heidelberg New York
London Paris Tokyo
Hong Kong Barcelona
Budapest

Volker ter Meulen
Institut für Virologie und Immunbiologie
der Universität Würzburg
Versbacher Strasse 7
97078 Würzburg, Germany

Martin A. Billeter
Institut für Molekularbiologie I
Universität Zürich
Hönggerberg
8093 Zürich, Switzerland

Cover Illustration: According to a drawing by Toni Cathomen and Roberto Cattaneo.

Cover design: Künkel und Lopka, Ilvesheim

ISBN-13: 978-3-642-78623-5 e-ISBN-13: 978-3-642-78621-1
DOI: 10.1007/978-3-642-78621-1

Production: PRODUserv Springer Produktions-Gesellschaft, Berlin

SPIN: 100 96 419 27/3020-5 4 3 2 1 0 – Printed on acid-free paper.

Preface

Measles, also called the greatest killer of children in history, still annually affects about 50 million individuals and causes close to a million deaths primarily in developing countries. Before the advent of measles vaccine some 30 years ago, these figures were roughly three times higher. Attenuated measles virus (MV) strains, all quite closely related to the original Edmonston isolate, have a very good record as a safe and highly efficacious vaccine and have brought down the measles toll in industrialized countries to almost negligible levels. However, recent outbreaks in the USA and Europe have again brought the measles problem to public attention. Sadly enough, these outbreaks were more instrumental in inducing activities to drastically reduce and hopefully finally eradicate measles than were the ten thousand times higher number of victims in developing countries. To reach this goal, as detailed in this volume, apparently it is not enough to rigorously enforce use of the existing vaccine as was the case with smallpox eradication: the intricacies of measles disease phenomena, in particular the generalized immune suppression which favors secondary infections, require more basic knowledge of the virus–host interactions and probably the development of new vaccines for special applications such as first immunizations of very young infants in developing countries.

In view of the increased research activities on measles encouraged by various health-related agencies, one important purpose of the present volume is to merge the large body of detailed knowledge on epidemiological, clinical and immunological aspects of MV accumulated during the past three decades with the more recent findings of the molecularly oriented sciences. It is to be hoped that this volume will catalyze exchange and collaborations between different disciplines and will facilitate the entry of newcomers to the MV field.

The appearance of a volume on MV is timely for other reasons, too. The MV genome and its components and products are now roughly defined and this is also true for cellular components crucial for MV replication, most notably the recently identified receptor molecules. Furthermore, cloning and expression techniques now facilitate biochemical investigations of MV, studies which until recently were reasonably feasible only with nonsegmented negative-strand RNA viruses (Mononegavirales) growing to high titers, such as vesicular stomatitis virus and Sendai virus. True reverse genetics involving the entire genome is still not feasible for any of these viruses for the time being, however, a good basis towards this goal has been elaborated by a number of laboratories: active transcription, replication and cellular transfer of diverse artificial subgenomic RNAs recovered from DNA constructs has been achieved, implicating genomic *cis*-acting elements of several representatives of the Mononegavirales, including MV, combined with reporter genes. These new tools will not only revolutionize the study of the different life styles of MV, including long-lasting persistence, and of the various disease phenomena triggered by MV, but will also be important in the elaboration of new generations of vaccines.

We wish to express our deep appreciation to all contributors for their spontaneous willingness to share their expertise in the production of a multifaceted and yet coherent volume.

Zürich and Würzburg

Martin A. Billeter
Volker ter Meulen

Addendum in Proof

In the meantime, the first rescue from cDNA of a representative of the order Mononegavirales, the very distantly related rabies virus, has been reported by the group of K.-K. Conzelmann (Tübingen), permitting application of the full power of reverse genetics to this virus. Using a rather different approach, rescue from cDNA has now been achieved for MV in the laboratory of M.A. Billeter (unpublished results).

List of Contents

List of Contributors

(Their addresses can be found at the beginning of their respective chapters.)

Clinical Spectrum of Measles

M. KATZ

1 Introduction

Acute measles, an exceptionally contagious viral infection, is a severe disease with a substantial degree of morbidity and a significant mortality (ROBBINS 1962; KRUGMAN et al. 1985). Before we had an effective vaccine, measles was an inevitable step in human development. Except for certain isolated areas of the world, all those who reached adulthood had been infected.

March of Dimes Birth Defects Foundation, 1275 Mamaroneck Avenue, White Plains, NY 10605, USA

1.1 History

Rhazes is credited with the first description of measles, in 840 AD, but there is no certainty that he distinguished it from smallpox (RHAZES cited in MAJOR 1954). This distinction is attributed to Sydenham, who states: "The measles generally attack children. On the first day they have chills and fever... On the second... cough... The nose and eyes run continually; and this is the surest sign of measles... [on] the fourth day... there appear on the face and forehead small red spots, very like the bites of fleas" (SYDENHAM 1850). Our modern understanding of the epidemiology of this infection is due to Panum, who in 1846 described the epidemic in the Faroe Islands and discerned both the length of the incubation period of measles and the establishment of a lifelong immunity in those who recovered from this infection (PANUM 1939).

1.2 Pathogenesis

The virus enters the host through the upper respiratory passages and the primary site of infection is respiratory epithelium (RILEY et al. 1978). The virus binds to a receptor, a cell surface glycoprotein, which is a membrane cofactor protein (CD46), a member of the regulators of complement activation gene cluster (NANICHE et al. 1993). The original impression, that the virus replicated extensively in the respiratory epithelium, has been put to question by extensive studies of live measles vaccine. These studies of the attenuated virus imply that the respiratory epithelium is not a major site of replication and that the local lymph nodes are (CHERRY 1992). Replication there leads to primary viremia and consequent infection of the reticuloendothelial system at multiple sites. Some 5–7 days after the initial infection, secondary viremia develops and leads to infection of the endothelial cells—causing enanthem—and epithelial cells—causing exanthem (RUCKLE and ROGER 1957). In the blood the virus is cell-associated with the leukocytes, particularly the monocytes (SALONEN et al. 1988).

2 Acute Measles

2.1 The "Classical" Disease

After an incubation period of approximately 10 days, during which time there is no outward sign of the disease, the infected child tends to develop a mild respiratory illness, easily confused with the common cold. A day or two later, these symptoms increase in intensity and now include fever, cough—which tends to become brassy, not unlike that characteristic of a viral croup—coryza, and conjunctivitis. Nausea is a common accompaniment.

This phase of measles corresponds to the development of an enanthem, which is responsible for a pathognomonic sign of measles, white to gray maculae with a pale red circumferential margin, visible on the buccal mucosa and the conjunctivae and known as Koplik spots (KOPLIK 1896). They appear 2 days before the rash. The underlying process of the enanthem is serous exudate and proliferation of lymphoid cells, mainly mononuclear ones, although some polymorphonuclear cells also appear, primarily around the capillaries. Characteristic giant multinucleated cells—the so-called Warthin-Finkeldey cells—resulting from fusion and containing several to more than 100 nuclei are seen in the lymphoid tissue (LIGHTFOOT et al. 1970).

Some 14 days after the exposure, or about 4 days after the onset of the initial symptoms, the characteristic exanthematous rash develops, whereas the enanthem subsides and Koplik spots disappear. The rash can be first seen on the forehead—typically at the hairline—and behind the ears, but over a period of 2 days it spreads throughout the body. The rash can be pale and subtle at its onset, but within a day or two becomes maculopapular and obviously red in color and, indeed, even somewhat hemorrhagic. By the third day it will have spread throughout the body. It begins to fade in the order of its original appearance, starting on the face, and can be expected to disappear by the tenth day. It leaves behind a brownish discoloration and even some desquamation, which soon heals and the skin reverts to its normal state. In the most severe cases, the rash can be quite intense and confluent. It can involve palms and feet and can be associated with edema, especially of the face.

There is fever, with the temperature usually reaching 40°C. Defervescence follows by lysis approximately 4 days after the onset of the rash; coryza and conjunctivitis also begin to subside at that time, but cough can persist for a week or so beyond the acute disease. Full recovery follows in most children, although asthenia lasting for a month or so is common.

In normal individuals, host defense develops and ultimately curbs the infection. Humoral immunity protects against a subsequent infection and is passively acquired through the placenta, tending to protect infants of seropositive mothers. This is a satisfactory level of protection in the industrialized countries, but in the developing countries, where exposure to measles during infancy is heavy this passive protection is usually inadequate after 9 months of age (CLEMENS et al. 1988).

Antibodies are first detected coincidentally with the appearance of the rash. They can be demonstrated by a variety of tests, viz. hemagglutination inhibition (HI), neutralization (N), complement fixation (CF), fluorescence tests (FA), enzyme-linked immunosorbent assay (ELISA), and immune precipitation (NEUMANN et al. 1985). HI is the most commonly used test, and a fourfold rise in the titer between the acute and convalescent serum confirms the diagnosis. The presence of this antibody correlates well with neutralization and indicates insusceptibility to measles (KRUGMAN et al. 1967). The first antibody to appear is of the IgM class, followed by the IgG1 and IgG4 classes (MATHIESEN et al. 1990). These antibodies correspond to the specific antigenic sites of the virus and they can be detected by immunoassays using purified viral proteins. Thus antibodies reacting with the N, or nucleocapsid, protein are present in the proportionally largest quantity, followed by those to the H, or hemagglutinin, protein, the F, or fusion, protein, and the M, or matrix, protein. The last appears in very small amounts.

Cell-mediated immunity is expressed in the CD4 and CD8 T cells, which helps to clear tissue-associated virus (van Binnendijk et al 1989, 1990).

2.2 Modified Measles

This term is applied to acute measles that has been attenuated by antibodies (Edmonson et al. 1990). Most commonly it develops in children exposed to measles who have received an immunoglobulin injection in an amount insufficient to abort the infection. Infants with transplacentally transmitted maternal immunoglobulin can also develop modified measles, when their passive immunity is on the decline and no longer fully protective. Modified measles seems to be also a rare consequence of a partial failure of measles vaccine. Whether this happens because the immunity induced by the vaccine has become less effective, or whether there was an insufficiently strong immune response to the original vaccine remains moot. In this form of modified disease, the children develop only an anamnestic IgG antibody response. A similar situation obtains in cases of recurrent measles, which are quite rare, but are well documented (Cherry et al. 1972).

The incubation period of modified measles tends to be longer than in classical measles and can extend up to 20 days after the exposure. On the other hand, the illness is much shorter, with the prodrome of 2 days and the disease lasting an additional 3 or 4 days. Moreover the clinical symptoms are much milder, characterized by a low grade fever, few or no Koplik spots, a minimal rash, and usually no adverse after effects.

3 Beyond the Core Disease

Although ordinary acute measles is a debilitating disease, responsible for significant absenteeism from school and corresponding loss of time from work by parents, it is its rarer expressions that cause major morbidity and mortality. The spectrum of these is remarkably wide. There is no general agreement whether they are part of the disease proper, or are its complications. In this discussion no such distinction will be made. Instead these expressions of measles will be categorized by the pathogenic process.

3.1 The Respiratory System

Respiratory symptoms and roentgenographic manifestations of pulmonary and hilar involvement are found in a majority of patients (KOHN et al. 1929), but serious clinical pneumonia—viral or bacterial—is rare, although its portent can be death (MASON et al. 1993). Laryngotracheobronchitis (LTB) tends to develop in the younger age group. During an epidemic nearly 19% of children experienced LTB. The mean age of this group was under 2 years, in comparison with the entire cohort, whose mean age was over 2 years (ROSS et al. 1992). Infants can experience bronchiolitis obliterans (OMAR and MANAN 1989; WIEBICKE and SEIDENBERG 1990).

3.2 The Heart

Myocarditis (FINKEL 1964), pericarditis and conduction defects can be detected by electrocardiography in a fifth of the patients but these abnormalities are transient and usually not associated with a clinical illness (ROSS 1952).

3.3 The Gastrointestinal System

Diarrheal disease is associated with vomiting during acute measles. The so-called measles enteritis is rare in the industrialized countries, but occurs with an observable frequency among measles patients under 2 years of age in the developing world (MONIF and HOOD 1970). Its consequences depend on the management of the dehydration, but in hospital cases no mortality has been attributed to this expression of measles. In a report from Indonesia more than a third of patients with measles who were admitted to the hospital had gastroenteritis, but this relatively high incidence reflects the fact that only the very sick are admitted. These children were 2 years of age, or younger, with none under 6 months of age (TAMBOEN et al. 1991).

3.4 The Hematopoietic System

The so-called black measles is a hemorrhagic form of this disease, which was common in the past in the industrialized countries, but today is observed mainly in the developing world. It has a sudden onset, presenting with high fever, followed by the development of a confluent rash, which becomes hemorrhagic; there is also bleeding from the mucous membranes. It is likely that this form of measles represents disseminated intravascular coagulopathy, but little is known about its pathogenesis.

On the other hand, thrombocytopenic purpura, which follows acute measles by several weeks, resembles the idiopathic thrombocytopenic purpura. Its prognosis is excellent (HUDSON et al. 1956).

3.5 The Central Nervous System

Although electroencephalographic abnormalities are found in half of all the patients with measles, acute clinical encephalitis is uncommon (AARLI 1974). It can present at any time from the prodrome to a week after the recovery. Its consequences can be quite severe, with mortality of 32% in one series of patients before 1947 and 12% in those who became ill after 1947 (LABOCETTA and TORNAY 1964). Symptoms are not unique to measles, but resemble those of any acute encephalitis. They include irritability, seizures, lethargy or coma, and ataxia. Cerebrospinal fluid (CSF) can show mononuclear pleocytosis and a slight elevation of protein, but in some cases it has been quite normal (MCLEAN et al. 1966). Although the pathogenetic process is most likely autoimmune (see chapter by Griffin), there have been reports of recovery of measles virus from the CSF (MCLEAN et al. 1966) and brain tissue of the affected patients (TER MEULEN et al. 1972).

Subacute sclerosing panencephalitis (SSPE) (TER MEULEN et al. 1983) is related to measles infection, but it can present many years after the acute disease. It is characterized by an insidious onset of a progressive cerebral dysfunction occurring over the course of weeks or months. The initial symptoms can involve alteration in personality and frequently include a deteriorating school performance. Myoclonic jerks, initially of the head and later also of the trunk and limbs, usually occur within 2 months. With progression of the illness, extrapyramidal dyskinesias including athetosis, chorea, ballismus and even dystonic posturing develop. Progressive loss of vision is due to focal chorioretinitis, cortical blindness, or optic atrophy. Ultimately the patient reaches a decorticate state and coma. SSPE is characterized by a slow progressive deterioration. The entire course of this virtually always fatal disease is variable, from weeks to years with periods of remissions. The average patient dies within 2 years, in rare cases patients have remained in a vegetative state for 10 years.

The clinical diagnosis is confirmed by the detection of serum measles antibodies in very high titers and by the presence of oligoclonal measles antibodies in the CSF. Characteristic EEG abnormalities are paroxysmal bursts of high voltage slow waves with associated spike discharges in two to three cycles per second, followed by a short period of flattened activity, the so-called burst-suppression pattern. The neuropathology is that of a subacute encephalitis accompanied by demyelination. Lesions involve the cerebral cortex, hippocampus, cerebellar cortex, basal ganglia, brain stem and spinal cord. Inclusions are seen within the nucleus and cytoplasm of the neurons and glial cells. Typically they consist of Cowdry type A homogenous eosinophilic material containing measles antigen detectable by fluorescent antibody staining. Ultrastructural studies reveal nucleocapsids of a paramyxovirus. The virus is defective in a variety of ways and has not been cultured directly from the brain tissue; it has been rescued by fusing the explanted brain cells with indicator cells. In one case it has also been rescued from a peripheral lymph node (HORTA-BARBOSA et al. 1971).

A somewhat less well defined form of measles encephalitis, inclusion body encephalitis, affects immunocompromised patients. It has been reported in those with leukemia, AIDS, and iatrogenic immunosuppression (AICARDI et al. 1977; AGAMANOLIS et al. 1979; HUGHES et al. 1993). It has a variable "incubation period", or the time interval from the recognized acute measles to the central nervous system disease, between 1 and 6 months. The clinical presentation is that of a nonspecific encephalitis, although convulsive seizures are almost invariable and occur early and epilepsia partialis continua was described in several cases. The outcome is usually death within 2 weeks; however, early treatment with ribavirin has been successful in one reported case (MUSTAFA et al. 1993). There is no pathological hallmark of this condition. It is a hemorrhagic leukoencephalitis with intranuclear inclusion bodies which contain paramyxovirus nucleocapsids. Polymerase chain reaction has been effective in demonstrating measles virus genome in the biopsied brain tissue (MUSTAFA et al. 1993).

4 Atypical Measles

Children who had received inactivated measles vaccine—no longer available—2–6 years later developed an unusually severe illness after an infection with the wild measles virus and, in rare instances, after immunization with the live measles vaccine (KARZON et al. 1965). The incubation period of atypical measles is the same as for the classical disease, but the prodrome is quite different. The disease is heralded by the sudden onset of high fever, reaching 40°C, headache, pleural and abdominal pain, and myalgia. Koplik spots are usually not seen. Although the rash appears in 2 or 3 days, as it does in classical measles, it has a distinct centripetal distribution; in some patients it is spotty, involving only wrists and ankles. It is a serious clinical disease, with respiratory distress, pneumonitis and pleuritis. Moreover, hepatosplenomegaly, paresthesias, and a profound asthenia are frequent. Neither coryza, nor conjunctivitis—so characterstic of measles—is a prominent feature. Of particular interest is the antibody response. In the CF and HI tests it tends to be tenfold higher than in classical measles.

5 Host Defense

During the acute disease and for several weeks thereafter, the patient experiences some degree of immune suppression. A positive tuberculin skin test reverts to negative, lymphoproliferative responses to mitogens are diminished, and production of lymphokines is abnormal. Clinical extent of this immune suppression can be variable. Some children develop secondary bacterial infections, such as otitis media and pneumonia and even tuberculosis (Starr and Berkovich 1965). Those with the nephrotic syndrome tend to have a remission of their disease. There is no definitive explanation for this suppressive effect in the face of a rather brisk immune response to measles, both humoral and cell-mediated. Griffin (this volume) has speculated that, because monocytes are infected by the virus, the immune defect probably results from an inadequacy of antigen presentation and production of IL1 and TNF-α, as these monocytes are damaged by the measles virus. In the absence of adequate production of interleukin-1 (IL1), lymphocytes do not proliferate and fail to produce significant amounts of IL 2.

Deficiency of vitamin A, even at a subclinical level contributes to the morbidity and even mortality of measles (Hussey and Klein 1990). Severity of clinical measles has been correlated in a direct proportion with the degree of depression of serum retinol concentration (Butler et al. 1993).

Measles in association with *Plasmodium falciparum* malaria apparently results in some degree of suppression of parasitemia. It shares this characteristic with influenza. Neither the clinical significance of this observation nor the mechanism underlying it is known (Rooth et al. 1992). Recent studies in Guinea-Bissau, Haiti, and Bangladesh have suggested that female children who received high-titer Edmonston-Zagreb live measles vaccine as infants experienced excess mortality as toddlers (see Clements and Cutts, this volume).

6 Conditions Putatively Associated with Measles

These are well defined diseases of uncertain etiology, which have had what could best be described as a "flirtation" with the measles virus. In each there is some suggestive evidence of the involvement of measles virus, but unlike SSPE, in which the virus has been isolated, it remains indirect.

6.1 Multiple Sclerosis

Since patients with multiple sclerosis tend to have an elevation of measles virus antibodies, the possibility that this virus is etiologically related has been raised. The problem with sorting out this hypothesis has been the ubiquity of measles infection and the presence of elevated antibodies to viruses other than measles (ter Meulen and Katz, in press). On epidemiological grounds, multiple sclerosis seems to have been absent in the isolated areas of the world free of measles and to have appeared in the wake of the introduced measles infection. On the other hand, an inability to isolate, or even identify in situ, the measles virus (Godec et al. 1992) has cast substantial doubt on this hypothesis.

6.2 Chronic Liver Disease

In a similar fashion, elevated measles antibodies have been reported in cases of chronic active hepatitis (CHRISTIE et al. 1987) and measles virus genome was identified in autoimmune chronic active hepatitis in 12 of 18 patients (ROBERTSON et al. 1987). This last finding, however, could not be reproduced in a study by KALLAND et al. (1989), who suggested that the presence of any measles "imprints" in such patients may be merely a reflection of their hyperimmune state. This is consonant with the findings of ANDJAPARIDZE et al. (1989), that RNA isolated from the peripheral blood lymphocytes of patients with chronic glomerulonephritis and systemic lupus erythematosus had, in many cases, specific sequences of measles virus nucleocapsid protein, whereas that from the controls did not.

Two types of acute liver disease in association with measles have been described in the literature and recently reviewed by KHATIB et al. (1993). Serum aminotransferase was elevated in 24 of 27 patients, all of whom were asymptomatic and a third of whom had atypical measles. Three cases of cholestatic jaundice were also described. With the respect to the latter, a number of potentially confounding factors were not evaluated.

6.3 Paget's Disease of the Bone

Ultrastructural studies of pagetic bone revealed structures resembling nucleocapsids of a paramyxovirus (REBEL et al. 1974). Thus, virologists tended to experience a dejà vu phenomenon recalling the SSPE story. A search for the virus began in earnest, and ultimately BASLE et al. (1986) detected measles virus RNA by in situ hybridization in bone of patients with this disease. In a subsequent study, GORDON et al. (1991) detected canine distemper virus sequences and none of the other paramyxoviruses, including measles. However, a year later, the same group of investigators, now using reverse transcribed RNA from pagetic bone, which was specifically amplified for distemper and measles sequences found evidence of canine distemper virus in most of their patients, but found both viruses in one patient and only measles virus in one other one (GORDON et al. 1992). In a similar study, CARTWRIGHT et al. (1993) also detected canine distemper virus sequences, but not those of measles, respiratory syncytial virus, and simian virus 5. There are no reports of an attempted isolation of the virus. In parallel with the human studies they also identified canine distemper virus transcripts in bone cells of dogs suffering with metaphyseal osteopathy (GORDON et al. 1993).

6.4 Otosclerosis

Even more tenuous evidence has been adduced for the measles virus etiology of this condition. Structures resembling nucleocapsids of a paramyxovirus have been seen in osteoblasts and preosteoblasts in stapes footplate fragments.

McKenna and Mills (1989) studied four temporal bone specimens from patients with otosclerosis and found reactivity with monoclonal antibodies against measles, especially in the area of active foci of the disease. However, Roald, et al. failed to detect any such evidence (Roald et al. 1992). There are no reports of an attempted isolation of the virus.

7 Conclusions

Although measles made its first appearance in the literature in 840 AD, and even as the seriousness of the acute disease has been recognized almost from the start, it is the last 30 years that have taught us about the enormous extent of its ravage. For some conditions we now have a large body of evidence implicating the measles virus; for other we have mere hints, or even just audacious guesses. Thanks to the available technology and the veritable explosion of molecular biological methods we can expect to learn more about measles in the next decade than we have learned in the preceding millennium.

References

Aarli JA (1974) Nervous complications of measles. Clinical manifestations and prognosis. Eur Neurol 12: 79–93

Agamanolis DP, Tan JS, Parker DL (1979) Immunosuppressive measles encephalitis in a patient with a renal transplant. Arch Neurol 36: 686–690

Aicardi J, Goutieres F, Arsenio-Nunes ML et al. (1977) Acute measles encephalitis in children with immunosuppression. Pediatrics 59: 232–239

Andjaparidze OG, Chaplygina NM, Bogomolova NN et al. (1989) Detection of measles virus genome in blood leukocytes of patients with certain autoimmune diseases. Arch Virol 105: 287–291

Basle MF, Fournier JG, Rozenblatt S (1986) Measles virus RNA detected in Paget's disease bone tissue by in situ hybridization. J Gen Virol 67: 907–913

Butler JC, Havens PL, Sowell AL et al. (1993). Measles severity and serum retinol (vitamin A) concentration among children in the United States. Pediatrics 91: 1176–1181

Cartwright EJ, Gordon MT, Freemont AJ et al. (1993) Paramyxoviruses and Paget's disease. J Med Virol 40: 133–141

Cherry JD (1992) Measles. In: Feigin RD, Cherry JD (eds) Pediatric infectious diseases, 3rd edn. Saunders, Philadelphia, p 1595

Cherry JD, Feigin RD, Lobes LA Jr. et al. (1972) Urban measles in the vaccine era: a clinical epidemiologic, and serologic study. J Pediatr 81: 217–230

Christie KE, Haukenes G (1987) Measles virus-specific IgM antibodies in sera from patients with chronic active hepatitis. J Med Virol 12: 267–272

Clemens JD, Stanton BF, Chakraborty J et al. (1988) Measles vaccination and childhood mortality in rural Bangladesh. Am Jour Epidemiol 128: 1330–1339

Edmonson MB, Addis DG, McPherson T et al. (1990) Mild measles and secondary vaccine failure during a sustained outbreak in a highly vaccinated population. J Am Med Assoc 263: 2467–2471

Finkel HE (1964) Measles myocarditis. Am Heart J 67: 679–683

Godec MS, Asher DM, Murray RS et al. (1992) Absence of measles, mumps, and rubella viruses in multiple sclerosis by polymerase chain reaction (PCR). Ann Neurol 32: 401–404

Gordon MT, Anderson DC, Sharpe PT (1991) Canine distemper virus localised in bone cells of patients with Paget's disease. Bone 12: 195–201

Gordon MT, Mee AP, Anderson DC et al. (1992) Canine distemper virus transcripts sequenced from pagetic bone. Bone Miner 19: 159–174

Gordon MT, May C, Bennett D et al. (1993) Canine distemper virus transcripts detected in the bone cells of dogs with metaphyseal osteopathy. Bone 14: 59–67

Horta-Barbosa L, Hamilton R, Wittig B et al. (1971) Subacute sclerosing panencephalitis: isolation of suppressed measles virus from lymph node biopsies. Science 173: 840–841

Hudson JB, Weinstein L, Chang T (1956) Thrombocytopenic purpura in measles. J Pediatr 48: 48–56

Hughes I, Jenney ME, Newton RW et al. (1993) Measles encephalitis during immunosuppressive therapy for acute lymphoblastic leukemia. Arch Dis Child 68: 775–778

Hussey G, Klein M (1990) A randomized trial of vitamin A in children with severe measles. N Engl J Med 323: 160–164

Kalland KH, Endresen C, Haukennes G et al. (1989) Measles-specific nucleotide sequences in autoimmune chronic active hepatitis (letter). Lancet 1: 1390–1391

Karzon DT, Rush D, Winkelstein W Jr. (1965) Immunization with inactivated measles virus vaccine: effect of booster dose and response to natural challenge. Pediatrics 36: 40–50

Khatib R, Siddiqui M, Abass M (1993) Measles associated hepatobiliary disease: an overview. Infection 21: 112–114

Kohn JL, Koriansky H (1929) Successive roentgenograms of the chest of children during measles. Am J Dis Child 38: 258–270

Koplik H (1896) The diagnosis of the invasion of measles from a study of exanthema as it appears on the buccal mucous membrane. Arch Pediatr 12: 918–920

Krugman S, Giles JP, Friedman H (1967) Studies on immunity to measles. PAHO / WHO Sci Publ 147: 353–360

Krugman S, Katz SL, Gershon AA, Wilfert C (1985) Measles (Rubeola). In: Infectious diseases of children, 8th edn. Mosby, St Louis, pp 152–166

LaBocetta AC, Tornay AS (1964) Measles encephalitis. Report of 61 cases. Am J Dis Child 107: 247–255

Lightfoot R, Nolan R, Franco M et al. (1970) Epithelial giant cells in measles as an aid in diagnosis. J Pediatr 77: 59–64

Major R (1954) A history of medicine. Thomas, Springfield, pp 238–239

Mason WH, Ross LA, Lanson J et al. (1993) Epidemic measles in the postvaccine era: evaluation of epidemiology, clinical presentation, and complications during an urban outbreak. Pediatr Infect Dis J 12: 42–28

Mathiesen T, Hammarstrom L, Fridell et al. (1990) Aberrant IgG subclass distribution to measles in healthy seropositive individuals, in patients with SSPE and in immunoglobulin-deficient patients. Clin Exp Immunol 80: 202–205

McKenna MJ, Mills BG (1989) Immunohistochemical evidence of measles virus antigen in active otosclerosis. Otolaryngol Head Neck Surg 101: 415–418

McLean DM, Best JM, Smith PA et al. (1966) Viral infections of Toronto children during 1965. II measles encephalitis and other complications. Can Med Assoc J 94: 905–910

Monif GR, Hood CI (1970) Ileocolitis associated with measles (rubeola). Am J Dis Child 102: 245–247

Mustafa MM, Weitman SD, Winick NJ et al. (1993) Subacute measles encephalitis in the young immunocompromised host: report of two cases diagnosed by polymerase chain reaction and treated with ribavirin and review of the literature. Clin Infect Dis 16: 654–660

Naniche D, Variro-Krishnan G, Cervoni F et al. (1993) Human membrane cofactor protein (CD46) acts as a cellular receptor for measles virus, J Virol 67: 6025–6032

Neumann PW, Weber JM, Jessamine AG et al. (1985) Comparison of measles antihemolysin test, enzyme-linked immunosorbent assay and hemagglutination inhibition test with neutralization test for determination of immune status. J Clin Microbiol 22: 296–298

Omar AH, Manan A (1989) Bronchiolitis obliterans in children—a report of six cases. Med J Malaysia. 44: 204–209

Panum PL (1939) Observations made during the epidemic of measles on Faroe Islands in the year 1846. Med Class 3: 829–886

Rebel A, Malkani K, Basle M et al. (1974) Ultrastructural characteristics of osteoclasts in Paget's disease. Rev Rhum Mal Osteoartic 41: 767–771

Riley EC, Murphy G, Riley RL (1978) Airborne spread of measles in a suburban elementary school. Am J Epidemiol 107: 421–432

Roald B, Storvold G, Mair IW et al. (1992) Respiratory tract viruses in otosclerotic lesions. An immunohistochemical study. Acta Otolaryngol (Stockh) 112: 334–338

Robbins FC (1962) Measles: clinical features. Pathogenesis, pathology, and complications. Am J Dis Child 103: 266–273

Robertson DAF, Zhang SL, Guy EC et al. (1987) Persistent measles virus genome in autoimmune chronic active hepatitis. Lancet 2: 9–11

Rooth IB, Bjorkman A (1992) Suppression of Plasmodium falciparum infections during concomitant measles or influenza but not during pertussis. Am J Trop Med Hyg 47: 675–681

Ross LJ (1952) Electrocardiographic findings in measles. Am J Dis Child 83: 282–291

Ross LA, Mason WH, Lanson J et al. (1992) Laryngotracheobronchitis as a complication of measles during an urban epidemic. J Pediatr 121: 511–515

Ruckle G, Rogers KD (1957) Studies with measles virus. II. Isolation of virus and immunological studies in persons who have had natural disease. J Immunol 78: 341–355

Salonen R, Ilonen J, Salmi A (1988) Measles virus infection of unstimulated blood mononuclear cells in vitro: antigen expression and virus production preferentially in monocytes. Clin Exper Immunol 71: 224–228

Starr S, Berkovich S (1965) The effect of measles, gamma globulin modified measles, and attenuated measles vaccine on the course of treated tuberculosis in children. Pediatrics 35: 97–102

Sydenham T (1850). The works of Thomas Sydenham, vol II (translated by RG Latham). Sydenham Society, London, p 250

Tamboen I, Sinuhaji AB, Sutantro AH (1991) Measles enteritis in infants and children. Paediatr Indones 31: 198–204

Ter Meulen V, Stephenson JR, Kreth HW (1983) Subacute sclerosing panenecphaltiis. In: Fraenkel-Conrat H, Wagner RR (eds), Compr Virol 18: 105–159

Ter Meulen V, Kackell Y, Muller D et al. (1972) Isolation of infectious measles virus in measles encephalitis. Lancet 2: 1172–1175

Ter Meulen V, Katz M (1994) An analysis of the proposed viral etiology of multiple sclerosis and related demyelinating diseases. In: Multiple sclerosis. Chapman and Hall, London (in press)

Van Binnendijk RS, Poelen MCM, Kuipers KC et al. (1990). The predominance of CD8+ T cells after infection with measles virus suggests a role for CD8+ class I MHC-restricted cytotoxic T lymphocytes (CTL) in recovery from measles. Clonal analyses of human CD8+ class I MHC-restricted CTL. J Immunol 144: 2394–2399

Van Binnendijk RS, Poelen MCM, deVries P et al. (1989) Measles virus specific human T cell clones. Characterization of specificity and function of CD4 helper/cytotoxic and CD8+ cytotoxic clones. J Immunol 142: 2847–2854

Wiebicke W, Seidenberg J (1990) Obliterating bronchiolitis following measles. Pneumologie 44: 1220–1222

The Epidemiology of Measles: Thirty Years of Vaccination

C.J. CLEMENTS[1] and F.T. CUTTS[2]

1 Introduction

Measles has been called the greatest killer of children in history. Over the centuries, it has been responsible for severe epidemics throughout the world. Although mortality rates from measles fell when socioeconomic conditions improved in developed countries, measles is still a major preventable cause of childhood mortality in developing countries. Measles vaccine was first licensed 30 years ago, and is one of the most cost-effective public health tools available. Through its use, most developed and many developing countries have reduced

[1] Expanded Programme on Immunization, World Health Organization, 1211 Geneva 27, Switzerland
[2] London School of Tropical Medicine and Hygiene, Keppel Street, London WC1E 7HT, UK

measles morbidity and mortality to a small fraction of preimmunization levels, and an estimated 1.54 million measles-associated deaths and over 90 million cases are prevented each year. Nonetheless, as of September 1994, the World Health Organization estimates that globally 45 million cases and 1.19 million deaths occur annually due to measles.

The eradication of smallpox from the world in 1977 demonstrated the feasibility of eradication of infectious disease. Public health officials, researchers, politicians and others have been systematically evaluating other candidates for potential eradication. In May 1988, the World Health Assembly selected poliomyelitis as another disease to be targeted for eradication. In 1990, international consensus was reached at the World Summit for Children to embark on a well-coordinated worldwide control programme as a major step towards the global eradication of measles in the longer term.

2 Measles-Associated Morbidity and Mortality

Measles ranks as one of the leading causes of childhood mortality in the world. In one community study in Kenya, measles accounted for 35% of reported deaths in infants up to 12 months of age, and for 40% of deaths in children 1–4 years old (Spencer et al. 1987). Complications affect multiple organ systems and are most common in the first 4–6 weeks after measles (acute complications) when disturbances of immune function are demonstrable. Measles also appears to

Table 1. Complications of measles

Acute complications
Diarrhoea
Pneumonia
Laryngotracheobronchitis
Otitis media
Malnutrition
Vitamin A deficiency (leading to corneal ulceration and blindness)
Stomatitis
Acute encephalitis
Later complications
Increased susceptibility to other infections
Subacute sclerosing Pan encephalitis (SSPE)
Complications of measles in pregnancy
Premature delivery
Foetal loss
Increased risk of maternal death
Possible (as yet unproven) association with measles
Multiple sclerosis
Osteosclerosis
Some autoimmune diseases

have more long-term effects on susceptibility to disease and risk of death (delayed complications), and has been postulated to be associated with autoimhave more long-term effects on susceptibility to disease and risk of death mune diseases in later life (Table 1). In immunocompromised hosts and in pregnant women, measles is particularly severe.

2.1 Acute Complications

In industrialized countries up to 10% of measles cases are associated with complications (ANONYMOUS 1990), the commonest being ear infection (around 5%), pneumonia (around 2.5%), diarrhoea (around 2.5%) and encephalitis (around 0.1%). In developing countries, the complication rate is much higher, and secondary infections are compounded by weight loss and functional micronutrient deficiencies. A community study from India showed 82% of children with measles had acute complications (NAIRN et al. 1989). A study of hospitalized children in Afghanistan (SRIVASTAVA et al. 1979) showed 30% of children with measles had diarrhoea, 84% had pneumonia, and 22% laryngotracheobronchitis (some had multiple complications).

Pneumonia may be caused directly by measles virus (particularly in malnourished or immunosuppressed children), (MITUS et al. 1959) by secondary infections with other viruses such as adenovirus and herpes simplex, (WARNER and MARSHALL 1976; KASCHULA et al. 1983) or by secondary bacterial infection. Measles is estimated to account for a median of 14% (range 6%–48%) of morbidity from acute lower respiratory tract infection (ALRI) and a median of 38% of ALRI mortality. (J. GINDLER et al., unpublished review 1993).

Mortality rates are higher in measles when cases are complicated by diarrhoea (KOSTER et al. 1981). A review of ten community-based studies in five developing countries showed that 15%–63% of measles cases were accompanied by diarrhoea. It was estimated that measles immunization can prevent 6%–26% of diarrhoea-associated deaths among children under 5 years of age (FEACHEM and KOBLINSKY 1983).

The combined effects of reduced nutrient intake, because of anorexia and mouth ulcers, and increased loss of water, minerals and protein into the gastrointestinal tract cause loss of more than 10% of body weight in over 25% of cases in developing countries (MORLEY et al. 1963; SCRIMSHAW et al. 1966) and measles may precipitate kwashiorkor (NETRASIRI and NETRASISI 1954). On average, children take 7 weeks to regain their former weight. Serum vitamin A and retinol binding protein levels are significantly depressed during the acute stage of measles, but usually revert to normal 8 weeks after recovery (REDDY et al. 1986). Measles commonly precipitates overt eye damage and blindness in countries where vitamin A deficiency is prevalent (INUA et al. 1983).

Encephalitis as a complication of measles usually occurs 7–10 days after appearance of the rash and at a relatively constant rate throughout different populations — around 0.1% of cases. The mortality rate for this complication is around 10%–30%.

Appropriate treatment of acute complications is likely to make a significant impact on measles mortality rates. Well proven management tools such as cotrimoxazole for respiratory infections, oral rehydration salts for dehydration, administration of vitamin A, and nutritional support are relatively low priced and should be made widely available.

2.2 Delayed Complications

Morbidity and nutritional status may remain significantly worse among children with measles than controls for several months after the acute infection. In a hospital-based study in India, at follow-up 6 months after discharge, measles patients reported more frequent episodes of infections than children without measles, the difference in number of days of illness being significant during the first 3 months (Bhaskaram et al. 1984). Community-based studies in India have shown significantly increased incidence of diarrhoea and cough, hospitalization for pneumonia (Reddy et al. 1986), and conjunctival dryness (Shahid et al. 1983) among affected individuals at 4–6 months after onset than controls.

A rare complication is subacute sclerosing panencephalitis (SSPE), which develops in between one and five per 100 000 cases and occurs an average of 6 years after the initial measles infection. SSPE is two to four times more common in males than females and is more common among children who acquired measles before 2 years of age. In industrialized countries, the incidence of SSPE has fallen dramatically with rising coverage of measles vaccine, indicating that the vaccine virus does not appear to cause SSPE (see chapter by Katz, this volume).

Measles remains a rare event in pregnancy in developing countries as most women of child-bearing age acquired measles at a young age. However, in industrialized countries (and it is to be predicted in developing countries in the future), the age distribution of measles cases is changed by immunization, resulting in measles infection in young adults. Atmar et al. (1992) described serious complications to the mother including pneumonitis, hepatitis, premature labour, and death. Adverse fetal outcomes include abortion and prematurity.

2.3 Case Fatality Rates

Mortality associated with measles varies between regions, countries, and even from one epidemic to another in the same locality. Measles acute case fatality rates (the ratio of the number of deaths occurring within 1 month of onset of measles rash to the total number of measles cases occurring) are currently estimated to be 0.1 %–0.2% in industrialized countries and around 50-fold higher in developing countries.

Quantifying case fatality rates (CFRs) is difficult and may be subject to bias, in that most information is usually obtained from sources which have an unrepresentative case load, such as hospitals. Outbreaks may also produce higher CFRs than are seen in endemic cases. For instance, an outbreak in Afghanistan resulted in a CFR of 28% for all ages and a CFR of 42% for children 0–4 years. The results from this study underline the added danger of acquiring measles at a young age (WAKEHAM 1978). Recognizing such difficulties, WHO/EPI reviewed community studies of CFRs and in the absence of country-specific information, now allocates CFRs based on this review of the literature (WHO 1991). Further work is needed, however, to refine how CFRs are measured.

Measles CFRs are higher among the young (GORDON et al. 1965, MCGREGOR 1964) unvaccinated (AABY et al. 1986), secondary cases (AABY et al. 1984; AABY and LEUWENBURG 1990; GARENNE and AABY 1990) low socioeconomic groups (BHUIYA et al. 1987), those lacking access to care including nutritional management (SCRIMSHAW et al. 1966) and vitamin A supplementation (BARCLAY et al. 1987; HUSSEY and KLEIN 1990), and those with underlying illness (DIAZ et al. 1992), particularly immunosuppressive diseases such as HIV/AIDS (KRASINSKI and BORKOWSKY 1989).

In developing countries, increased case fatality among secondary and subsequent generations of cases has been documented after controlling for age and nutritional status. This may arise not only because of higher infectious load of measles virus, but also because of a higher probability of secondary infection or larger dose of superinfecting pathogens. Case fatality is therefore likely to be highest when there is a high proportion of secondary cases, such as in outbreaks, and in crowded urban areas, particularly if there is poor medical and nutritional care. In such conditions the CFR can be as high as 20%. In contrast, in the USA, SUTTER et al. (1991) found no difference in severity between primary and secondary cases in a very different sociological setting where the overall severity of measles was much less and mortality was minimal.

Studies on the effect of preexisting malnutrition have yielded inconsistent results. There are many biological reasons to expect more severe measles in malnourished children, who have depressed cellular immune function and often micronutrient deficiency (GREENWOOD and WHITTLE 1981). Some studies have reported increased severity of measles and higher case fatality among malnourished children (MORLEY 1973; SAVAGE 1967). However, other studies have shown no effect of nutritional status on mortality (KOSTER et al. 1981), and nutritional status may be confounded by intensity of exposure. Multiple cases are more likely to occur in crowded households with large families, which are also those at risk of malnutrition (AABY 1988).

As with morbidity, mortality may be increased for several months after measles infection (AABY and CLEMENTS 1989; HULL et al. 1983). Death may result from a variety of apparently unrelated conditions in the months following measles infection. One possible explanation of these observations is an altered cell-mediated immune response as a result of measles infection. It has been postulated that persistence of measles virus may account for increased delayed mortality, but there is no empirical evidence for this. Proof accumulates, however, for some form of immune suppression (see chapter by Griffin, this volume).

The younger the age of infection, the greater the effect of this delayed mortality phenomenon, emphasizing the urgent need to protect children by immunization at the youngest possible age. It is thought that the true impact of measles virus infection is much greater than is actually recorded in the number of deaths generally attributed directly to measles, making the control of this disease an even greater priority. To date, studies describing delayed mortality have all been conducted in Africa. It is important to confirm these observations in Asia and the Americas.

3 Measles Epidemiology

Measles virus is most transmissible from a 4-5 days prior to the onset of rash to 1–2 days after rash onset. It is highly infectious: one study estimated that 76% of household exposures of susceptibles led to measles transmission, compared to 61% for varicella and only 31% for mumps (Hope Simpson 1952).

Prior to measles immunization, immunity to measles was acquired by natural infection or by passive immunization of neonates by maternal antibody transferred in utero to the fetus. This passive immunity wanes gradually as the growing infant catabolises maternal antibodies. The majority of infants are susceptible to measles by the age of 6–9 months in developing countries and 9–12 months in industrialized countries. Because measles transmission requires contact between an infectious individual and a susceptible individual, the epidemiology of measles is determined by the likelihood of such contact in different population groups.

In large unimmunized populations, measles epidemics occur at regular intervals. In a given population, an increase in the number of persons susceptible to infection or in the frequency of contact between infectious individuals and susceptibles will lead to an increase in measles cases. Measles incidence rises, thus reducing the number of persons susceptible to measles, until the chance of an infectious person making contact with a susceptible person decreases, when measles incidence begins to fall. As measles incidence falls, the number of susceptibles rises again, until a "threshold density" of susceptibles is passed, at which point measles transmission increases and incidence rises once more. A classical outbreak in an unimmunized population in India was described by Nairn et al. (1989) (Fig. 1), when measles spread rapidly through the village population.

New susceptibles may enter a population by birth or by in-migration of older persons who have not been exposed previously. In large populations, the rate of entry of new susceptibles is high enough to sustain continuous measles transmission, so that measles is endemic, though still demonstrating periodic cycles of epidemics. In small, isolated populations, measles dies out between epidemics, and susceptibles accumulate until a new infectious individual enters the area, when an explosive epidemic may occur. A famous outbreak in the Faroe Islands followed 65 years after the epidemic of 1781 (Panum 1939), affecting all age groups born since the previous epidemic. In rural areas which have high population movement to and from surrounding areas such as towns or trading

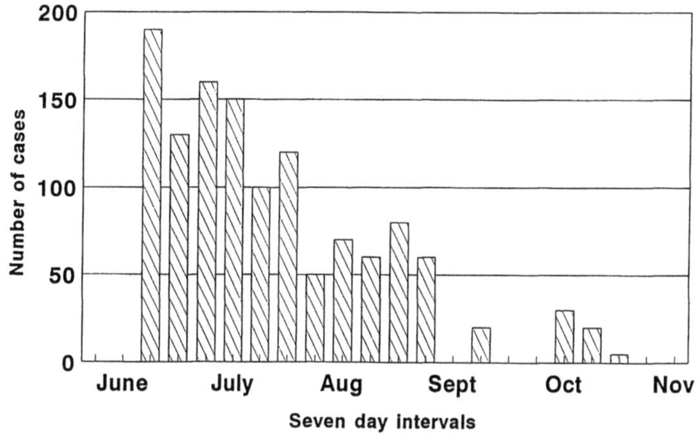

Fig. 1. Spread of measles during an outbreak in an unimmunized population. Cases reported from an outbreak of measles in Pikhi District, India, 1986. (From NAIRN et al. 1989)

centers, measles epidemiology may resemble that in urban areas, with frequent introduction of measles causing small clusters of cases and shorter disease-free periods (VOORHOEVE et al. 1977).

There appears to be an inter-relationship between continuous urban transmission and rural outbreaks. BARTLET (1957) and BLACK (1966) suggested that, in an isolated situation, a minimum population of 400 000–500 000 susceptible individuals is required to sustain measles transmission. If this is correct, rural transmission is likely to be initiated by cases from nearby cities. This "seeding" phenomenon was documented in an outbreak in Oman (ANONYMOUS 1993) when imported cases to one village resulted first in a surge of cases in nearby villages followed by a larger outbreak in more distant villages (Fig. 2). This observation underlines the importance of improving coverage not only at the country level but even at the village and community levels as well. Improved surveillance and special studies are needed to obtain better documentation of the foci of measles transmission in countries with high coverage.

The average age at measles infection depends on the age at which maternal antibody is lost and on age-related changes in frequency of contact with other persons. Measles occurs among younger children in developing than industrialized countries because of the higher birth rates, more crowded living conditions, and earlier loss of maternal antibody in the former (BLACK et al. 1986). Similarly, the age at infection is earlier in urban than rural areas. The average age of infection was 4–6 years in most industrialized countries in the early 1960s, prior to the introduction of widespread immunization (WALSH 1983). In Senegal, 100% of urban children were seropositive at 18–24 months of age, whereas in rural areas 100% seropositivity was not reached until 10 years of age (BOUE 1964).

Cultural factors which affect the likelihood of an infected person coming into contact with a susceptible individual may affect the seasonality of measles. In the UK, for example, FINE and CLARKSON (1982) found that changes in measles incidence were related to changes in the chance of contact between infectious and susceptible persons when school terms began and ended. Measles transmission fell during school holidays, and rose after school opening.

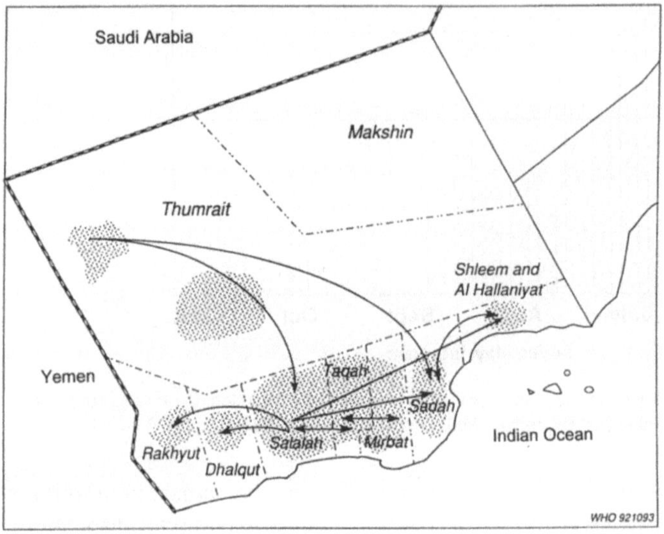

Fig. 2. Dispersal of measles cases in Dhofar Region, Oman, December 10, 1991 to May 31, 1992, (ANONYMOUS 1993)

4 Measles Vaccines

Measles virus was isolated by ENDERS and PEEBLES in 1954. Both live and killed vaccines were developed initially, and, by the end of 1950s, Enders and colleagues had developed the Edmonston B strain of live attenuated measles vaccine (ENDERS 1962). Because of its reactogenicity, gamma globulin was often administered simultaneously (KRUGMAN et al. 1963). By the middle-late 1960s, new strains of measles vaccine were developed in the USA, Japan, Yugoslavia, the USSR and China, by further attenuation of Edmonston (AIK-C), Edmonston A (Schwarz), Edmonston B (Moraten, Edmonston Zagreb) or separate isolates (Leningrad 16, CAM-70). Further attenuated vaccines were less reactogenic and were suitable for widespread use without concurrent administration of gamma globulin.

One of the first measles vaccines was a formalin-inactivated vaccine derived from the Edmonston strain. Usually, three doses of inactivated vaccine or two doses of inactivated and one dose of live vaccine were administered at monthly intervals (KRUGMAN et al. 1965). Use of inactivated vaccine was stopped in 1967 when it was realized that immunity was short-lived and that recipients were at risk of atypical measles on exposure to live measles virus (CENTERS FOR DISEASE CONTROL 1967).

Measles vaccines are heat stable in the freeze-dried condition. They survive well at temperatures between 0°C and + 8°C in the refrigerator, but can be taken out of refrigeration for a considerable time before there is serious loss of potency. Once reconstituted, however, at 22°–25°C, they suffer approximately 50% loss in potency in 1 h, and at temperatures over 37°C, they are inactivated within 1 h.

It is therefore extremely important to keep reconstituted vaccines cool and protected from sunlight. It is anticipated that measles vaccine vials will shortly be provided with heat-sensitive colour indicators. Such colour indicators will not replace cold chain monitors; however, they will allow health workers to know when the vaccine which has not yet been reconstituted should no longer be used. Reconstituted vials of measles vaccine are discarded at the end of the working day.

4.1 Immunogenicity

Measles vaccine is highly immunogenic when potent vaccine is given in the correct dose to children of appropriate ages. However, the presence of maternal antibody markedly reduces seroconversion rates. Where measles among young infants is not a public health concern, it is recommended to administer vaccine at 12–15 months of age, a time when the majority of children no longer have maternal antibody circulating and seroconversion rates will be very high. Where measles is occurring earlier—the case in most developing countries—immunization is recommended at 9 months of age. This is a balance between immunizing later in life (when no maternal antibody remains and seroconversion will be high), but the infant may be exposed to measles in the mean time, and immunizing early in life before the infant is exposed to measles (when seroconversion will be low).

4.2 Vaccine Efficacy

Vaccine efficacy is typically determined by comparing the rate of measles in immunized children with that of measles in unimmunized children. Under most field conditions, the vaccine efficacy is thought to be at least 85% for children receiving the vaccine at 9 months of age. Vaccine efficacy increases to 95% or higher when potent vaccine is administered in the absence of maternal antibodies at 12–15 months of age.

4.3 Duration of Immunity

Infection with the wild measles virus confers life-long immunity. The measles vaccine virus is attenuated and antibody levels after immunization are lower than those after disease. Nonetheless, in most individuals, immunity induced by standard measles vaccine (administered at an age when maternal antibody has been lost) lasts at least 21 years (Miller 1987). Antibody persists longer when there is boosting from exposure to circulating wild virus (Krugman 1983; Zhang and Su 1983). However, even in isolated communities, antibodies have been shown to persist for at least 16 years after immunization (Xiang and Chen 1983).

When measles antibody falls to low levels, reexposure to measles virus (wild or vaccine virus) stimulates a secondary immune response, in which IgG levels rise rapidly and peak approximately 12 days after reinfection. Although in the majority of these persons, reinfection by wild virus will cause only a subclinical boost of antibody levels, cases of clinical measles have been documented in persons who had originally seroconverted after immunization (secondary vaccine failures) (REYES et al. 1987; MATHIAS et al. 1989). If antibody levels are high prior to exposure, reinfection is prevented and a boost is rarely seen (ZHUJI MEASLES VACCINE STUDY GROUP 1987). As outbreaks of measles become rare, exposure to the wild virus will diminish, and therefore this boosting effect by the wild virus will diminish also. The duration of vaccine-induced immunity in such circumstances needs further study.

4.4 Safety of Standard Vaccines

Standard dose further attenuated measles vaccines are among the safest vaccines in use today (FINE 1993). Local reactions at the site of infection are minimal. The main reactions are associated with a mild, measles-like syndrome which occurs in 2%–3% of recipients approximately 1 week after immunization. Febrile convulsions may occur in approximately 1 per 1000 vaccinees, but generally resolve without sequelae. Encephalitis or encephalopathy have rarely been reported after measles immunization (less than one per 100 000 vaccinees, compared with an incidence of 5–40 per 10 000 measles cases). There are some groups who are at increased risk of adverse events after measles vaccines. Chief among these are children who are immunosuppressed (e.g. congenital immune deficiency, leukaemia, lymphoma or generalized malignancy, or those on high dose steroids); measles vaccine is contraindicated in these children. In contrast, children infected with HIV have been shown to have no increase in adverse events compared with controls (VON REYN et al. 1987). HIV infected children are in particular need of measles vaccine because of the severity of measles disease in such children.

4.5 Alternative Routes of Administration

Initial replication of wild measles virus occurs on the mucosa of the respiratory passages and the conjunctiva. It is therefore appealing to attempt introduction of the vaccine virus in the same way, thereby allowing replication without the interference of circulating maternal antibody. Nonparenteral routes of administration of vaccine (such as intranasal, aerosol) may be more efficient in stimulating the formation of secretory IgA to provide local immunity against reinfection (OGRA et al. 1980).

Administration of vaccine by aerosol has given equivalent seroconversion rates to the subcutaneous route in most studies (SABIN et al. 1983; WHITTLE et al. 1984). However, this route has proved technically difficult and impractical (KHANUN et al. 1987), and there have been concerns about cross-contamination with masks (WHITTLE et al. 1984). While the aerosol route has found favour in some mass campaigns in the Americas, it is still not widely accepted. However, this low cost route remains an attractive potential alternative which requires further research.

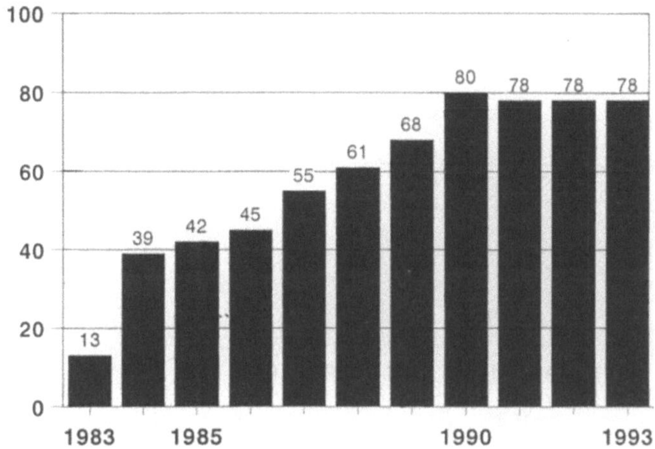

Fig. 3. Global coverage with measles vaccine 1983–1993. (From information supplied to WHO)

5 Current Progress in Measles Control

5.1 Impact of Vaccination

The use of measles vaccine has had a major impact on the epidemiology of measles. Before the vaccine became available, virtually all children contracted measles, thus an estimated 130 million cases and around 7–8 million deaths occurred globally each year (assuming a CFR of 7% in the pre-immunization era in developing countries). The activities of governments, the Expanded Programme on Immunization (EPI), and collaborating agencies, have resulted in a dramatic increase in coverage which has contributed significantly to reducing both measles morbidity and mortality. By 1991, approximately 80% of the world's children under 1 year of age were reported to have received the vaccine (Fig. 3), and it was estimated that over 1 million deaths were prevented annually.

Despite impressive progress to date in immunizing the world's children, WHO estimates that, as of September 1994, 45 million cases and around 1.19 million deaths from measles occur annually throughout the world. Thus measles is still responsible for more deaths than any other EPI target disease. Reported measles cases have declined since the introduction of the first measles vaccines in the 1960s (Fig. 4). Because only 5% or less of actual cases are reported in many situations, the actual numbers cannot be taken at truly representative. Nonetheless the global trend is clear; as measles vaccine coverage has risen, the number of cases has declined.

In locations of sustained high vaccine coverage and where there is good surveillance, a marked reduction in measles cases has been noted. Several countries in eastern and southern Africa have sustained coverage of 70% or more and reduced mortality and morbidity from measles to a small fraction of previous levels. In most countries in the Arabian Peninsula, coverage is over 80%

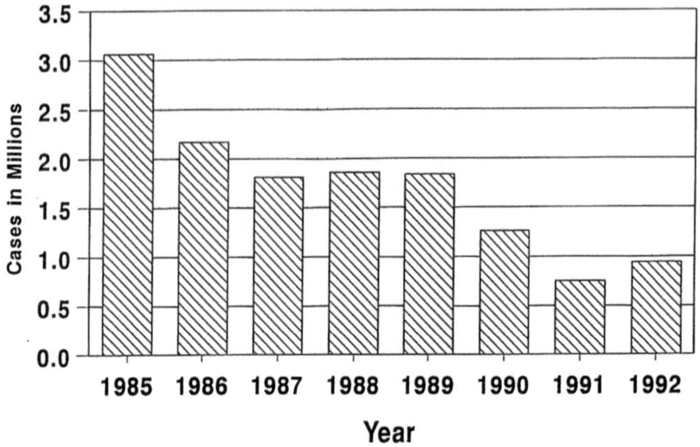

Fig. 4. Number of cases of measles reported globally 1985–1992. (From information provided to WHO)

and measles has virtually disappeared. In Malaysia, coverage is 79% and only around 400 cases were reported in 1992, a thirtieth of the number reported 10 years ago.

Urban slums represent a major challenge for measles control, as families often use health services less frequently so that coverage is low, and measles is likely to be more severe when it occurs. Nonetheless, some African cities have demonstrated the feasibility of controlling measles in these environments. Maputo city, Mozambique, for instance, has reduced reported measles incidence rates by 92% and mortality by 96% from preimmunization levels by achieving and sustaining coverage of over 90% of children in the first year of life (Cutts et al. 1994). Measles incidence has fallen in all age groups, including children under 9 months of age, and there has been a shift to older ages. Kinshasa, Zaire, has achieved similar reductions in measles incidence by attaining close to 90% coverage with "medium" titre Edmonston Zagreb (EZ) measles vaccine at the age of 6 months (Cutts et al. 1994). Reported incidence rates decreased by over 90% from 37.5/10 000 in 1980 to 6.5/10 000 in 1990 and 1.6/10 000 in 1991. Reported measles incidence declined in all age groups, but the decrease was greatest among infants under 9 months of age; 32% of cases were below 9 months in 1986–1987 before introduction of EZ vaccine. This dropped to 23% of cases in 1990–1991 after introduction of EZ vaccine. At the same time, there was a relative increase in cases among children over 23 months of age from 29% of cases in 1986–1987 to 43% in 1990–1991.

5.2 Age of Affected Individuals

When assessing changes in measles age-specific incidence, it is important to distinguish between proportions, relative frequencies and absolute incidence. Following a period of successful immunization in a community, two events commonly occur: (1) an increase in the proportion of measles cases which occur in immunized children, and (2) an increase in the proportions of children who acquire measles before the scheduled age of immunization or after the upper limit of the age group targeted for immunization.

As coverage increases, the average age of cases tends to rise. Achievement of high vaccine coverage induces a period of low incidence, termed the "honeymoon period" (McLean and Anderson 1988). During this period, cohorts of susceptible children who were not immunized in the early years of the programme can reach older ages before being exposed. Susceptibles gradually accumulate until the epidemic threshold is reached, when "posthoneymoon" outbreaks occur (see below), which affect predominantly children of older ages. Though the *proportion* of cases in older children increases after immunization, the absolute *number* of such cases may fall due to the overall reduced incidence rate of measles.

High coverage of children in the 9–11 month target age group soon has an impact on incidence among children from 9 months onwards. However, the proportion of cases occurring prior to 9 months may increase, though the actual number of cases may not rise. In urban areas of developing countries, some studies have shown that up to a third of reported cases occur before 9 months of age when coverage is moderately high (Taylor et al. 1988). These data have served as a critical impetus for researchers to find a measles vaccine which could successfully immunize infants earlier in life. Over the next decade, it is hoped that a vaccine will be developed which can be administered early in life, thus protecting infants from attacks of measles before the scheduled age of administration of presently available vaccines, and facilitating the attainment of high coverage.

5.3 Outbreaks

Even in countries placing additional efforts behind measles control, outbreaks of measles are expected to continue and do not necessarily indicate programme failure. The commonest cause for outbreaks in developing countries is still a failure to immunize (low vaccine coverage). But, for a number of reasons, outbreaks have occurred even in developed and industrialized countries which have had high measles vaccine coverage. First, even when coverage is moderately high, susceptibles gradually accumulate over a number of years until a "posthoneymoon" outbreak occurs. Second, there may be pockets of low coverage, where outbreaks can occur after measles is introduced. Third, outbreaks have occurred among the small percentage of immunized persons who do not respond to the vaccine even when it is administered under optimal conditions. Lastly, outbreaks may occur in immunized populations if there has been a problem with vaccine handling, so that vaccine efficacy is low.

Typical "posthoneymoon" outbreaks were reported in 1988 in Harare, Zimbabwe, when reported coverage was 83% (Kambarami et al. 1991), in Muyinga sector, Burundi, when coverage was 64% (Chen RT et al. 1994), and in Hong Kong when coverage was 80% (Yu-Lung Lau et al. 1992) (Fig. 5).

Yoshioka Y, Fujita Y, Ohtsubo E (1990) Nucleotide sequence of the promoter distal region of the tra operon including traI (DNA helicase I) and *traD* genes of plasmid R100. J Mol Biol 214: 39–53
Zerbib D, Jakowec M, Prentki P, Galas DJ, Chandler M (1987) Expression of proteins essential for IS1 transposition: specific binding of InsA to the ends of IS1. EMBO J 6: 3163–3169
Zerbib D, Polard P, Escoubas JM, Galas D, Chandler M (1990a) The regulatory role of the IS1-encoded InsA protein in transposition. Mol Microbiol 4: 471–477
Zerbib D, Prentki P, Gamas P, Freund E, Galas DJ, Chandler M (1990b) Functional organization of the ends of IS1: specific binding for an IS1-encoded protein. Mol Microbiol 4: 1477–1486
Zuber U, Schumann W (1993) The eighth copy of IS1 in *Escherichia coli* W3110 maps at 49.6 min. J Bacteriol 175: 1552
Zuerner RL (1994) Nucleotide sequence analysis of IS1533 from *Leptospira borgpetersenii*: identification and expression of two IS-encoded proteins. Plasmid 31: 1–11

attenders had demonstrable serological antibodies to measles. Thus, even with a vaccine efficacy of over 95%, outbreaks can occur among vaccine failures under conditions of close contact such as school settings, although the overall attack rates will usually be low.

A large outbreak in Hungary in 1988–1989 was thought to reflect low vaccine efficacy (AGOCS et al. 1992). The immunization status was known for 12 890 (76%) of cases in 1989. Of these, 8006 (62%) occurred in immunized persons. Of the 14 537 (86%) with a known year of birth, 8608 (59%) were born during 1967–1973 (16–22 years of age). Persons born during 1971 and 1972 had the highest attack rates, suggesting that vaccine efficacy had been low when vaccine was administered in the mass campaigns during the beginning of the vaccination programme in that country. This emphasizes the importance of monitoring the quality of vaccination practices in addition to coverage.

Such examples of outbreaks occurring in relatively highly immunized populations do not weaken the argument for attaining high coverage, which is likely to lengthen the period between epidemics. With control programmes and effective surveillance systems in place, the size of expected outbreaks can also be minimized, and it is important to note that measles incidence has been greatly reduced from that of the preimmunization era.

5.4 Herd Immunity and Elimination

Mathematical models of infectious diseases have shown that once a population reaches a certain level of immunity, the remaining unimmunized individuals will be protected by the immunity of the surrounding "herd" (ANDERSON and MAY 1983). But due to the highly infectious nature of the measles virus, the required level is thought to be very high. It is also recognized that the nonrandom distribution of the unimmunized in a population will make it difficult to predict the effects of herd immunity.

It appears that the level of coverage needed to interrupt transmission will vary with the epidemiological situation. A low-density rural or island population with few imported cases may need much lower levels of coverage to achieve herd immunity, whereas it has been demonstrated that coverage of 83% and a vaccine efficacy of around 85% are not sufficient for a city like Harare (KAMBARAMI et al. 1991). In crowded conditions, it is likely that even 95% vaccine coverage using a vaccine whose efficacy is less than 90% may not be sufficient to eliminate measles.

6 New Schedules, New Challenges

The world is no longer satisfied with just controlling measles. Even with the limitations of the available vaccines and strategies, some countries are aiming to eliminate the disease. An added impetus has been provided by the World Health

Assembly and the 1990 World Summit for Children, which set specific goals of case and death reduction by 1995 as a step towards global eradication. The fundamental activity undertaken at a country level to control measles is to raise coverage with one dose of standard measles vaccine as high as possible. Several complementary strategies are now planned or operational to reduce overall cases and deaths, to reduce cases and deaths in young infants, or to eliminate measles virus from a country or subregion. Some of these involve making better or more strategic use of what is already available, while others involve totally new technologies.

A number of high or middle income countries have introduced an additional routine dose of vaccine. Many European countries and the United States give the second dose to school age children, often including it as a multi-valent vaccine such as MMR (mumps, measles, rubella). This two dose schedule aims to reduce primary vaccine failures. Measles has virtually disappeared from Sweden since introduction of a two dose MMR schedule and attainment of high coverage with both doses (Bottiger et al. 1987), and is now an uncommon infection of much of Europe and North America.

Campaigns have been used successfully in the region of the Americas to provide either a missed first dose or an additional dose of vaccine to all persons aged 9 months to 15 years. In the 2 years ending 1993, most countries in Latin America had undertaken measles mass campaigns which included all children 9 months to 15 years of age. The results so far indicate measles has virtually ceased to circulate in these countries. Until measles is eradicated, it may be necessary, however, to repeat campaigns to maintain the virtual absence of cases, otherwise numbers of susceptibles will accumulate until an outbreak occurs.

6.1 Future Directions and Needs

As described earlier, the continued occurrence of cases before the age of immunization has been a constant concern to immunization programmes. One solution may be to raise coverage in older eligible children, so reducing the amount of virus circulating and the absolute number of children in the pool of susceptibles. However, over the last 10 years, considerable effort has been placed in trying to develop vaccines which can be administered at an earlier age.

While all measles vaccine strains are thought to be equally immunogenic in older infants and children without maternal antibodies, some vaccine strains, in particular EZ (Markowitz et al. 1990; Job et al. 1991) and AIK-C (Tidjani et al. 1989) strains, are more immunogenic in the presence of maternal antibody. Furthermore, increasing the potency of live attenuated vaccines was found to increase their immunogenicity in the presence of maternal antibody (Whittle et al. 1988).

In 1989 the World Health Organization recommended the use of high titer EZ vaccine for use in children aged 6 months in countries in which measles before the age of 9 months is a significant cause of death. However, before the high titer vaccine became available for widespread use, concerns were raised about its long term safety. In June 1992, a panel of experts reviewed data from several studies and recommended that: "High titre (equal to or greater than log 4.7 infectious units per human dose as expressed in relation to the International Reference Reagent for measles vaccine) measles vaccine derived from the original Edmonston measles virus isolate should no longer be recommended for use in immunization programmes" (ANONYMOUS 1992).

While efforts during the 1980s were directed at defining the differences between live attenuated vaccine strains which may have permitted immunization in the presence of maternal antibody, more recent approaches have focused on the use of new molecular techniques to design vaccines. Among the approaches being investigated, three approaches to immunizing before 6 months of age look particularly promising. The insertion of internal proteins of the measles virus into the bacillus Calmette Guerin (BCG) may allow priming of infants in the first few weeks of life. Thus if wild measles virus were encountered, or a later dose of standard vaccine given, a booster effect would result in either no infection or a mild infection.

The second possibility is the development of a recombinant vaccine based on the canary pox virus, an organism which cannot replicate in humans. The third is the use of an immune stimulating complex (ISCOM) to present measles proteins to the immune system. All three developments represent new technologies which may, over the next decade, provide radical new approaches to measles control.

The supply of standard vaccine is critical if programmes are to continue their work of immunizing children. There is adequate manufacturing capacity for measles vaccine to meet the current global needs. However, to meet a significant increase in demand, industry would need to expand capacity—requiring a major investment in equipment. There would also be roughly a 2 year lead time before new plant could be operational. Global vaccine requirements have been increasing as countries have adopted more vaccine-intensive strategies for controlling measles such as multiple dose schedules, outbreak response, national vaccination days or campaigns.

It has become increasingly difficult to predict accurately what global vaccine needs will be over the next decade, but a trend can be anticipated where more countries are likely to administer more than one dose per child. Evaluation of new measles control strategies must include a consideration of the impact on vaccine demand and a commitment to assure the supply and financing of the vaccine.

7 Summary

Measles is a highly infectious disease which has a major impact on child survival, particularly in developing countries. The importance of understanding the epidemiology of this disease is underlined by its ability to change rapidly in the face of increasing immunization coverage. Much is still to be learned about measles epidemiology and the best strategies for administering measles vaccines, as well as about the biological mechanisms of action of measles vaccines. However, it is clear that tremendous progress can be made in preventing death and disease from measles with existing knowledge about the disease, and by using the presently available vaccines and applying well tried methods of treating cases. Research in the coming decade may provide improved strategies and more effective vaccines for use in immunization programmes.

Acknowledgments. The authors wish to acknowledge the assistance of Drs. LE Markowitz and NA Halsey in preparing this manuscript.

References

Aaby P (1988) Malnutrition and overcrowding/intensive exposure in severe measles infection: review of community studies. Rev Infect Dis 10: 478–491

Aaby P, Clements CJ (1989) Measles immunization research: a review. Bull WHO 67: 443–448

Aaby P, Leuwenburg J (1990) Patterns of transmission and severity of measles infection: a reanalysis of data from the Machakos area, Kenya. J Infect Dis 161: 171–174

Aaby P, Bukh J, Lisse IM, Smits AJ (1984) Overcrowding and intensive exposure as determinants of measles mortality. Am J Epidemiol 120: 49–63

Aaby P, Bukh J, Leerhoy J, Lisse IM, Mordhorst CH, Pederson IR (1986) Vaccinated children get milder measles infection: a community study from Guinea-Bissau. J Infect Dis 154: 854–863

Agocs MM, Markowitz LE, Straub I, Domok I (1992) The 1988-1989 measles epidemic in Hungary: assessment of vaccine failure. Int J Epidemiol 21: 1007–1013

Anderson RM, May RM (1983) Vaccination against rubella and measles: quantitative investigations of different policies. J Hyg (Camb) 90: 259–325

Anonymous (1989) Measles – United States, first 26 weeks, 1988. Morbid Mortal Wkly Rep 38: 601–605

Anonymous (1990) Measles – United States, first 26 weeks, 1989. Morbid Mortal Wkly Rep 38: 863–871

Anonymous (1992) Safety of high titer measles vaccine. Wkly Epidemiol Rec 67: 357–362

Anonymous (1993) Measles outbreak, Oman. Wkly Epidemiol Rec 7 (68): 45–47

Atmar RL, Englund JA, Hammil H (1992) Complications of measles in pregnancy. Clin Infect Dis 14: 217–226

Barclay AJG, Foster A, Sommer A (1987) Vitamin A supplements and mortality related to measles: a randomized clinical trial. Br Med J 294: 294–296

Bartlet MS (1957) Measles periodicity and community size. JR Stat Soc A 120: 48–70

Bhaskaram P, Reddy V, Raj S, Bhatnagar RC (1984) Effect of measles on the nutritional status of preschool children. J Trop Med Hyg 87: 21–25

Bhuiya A, Wojtyniak B, D'Souza S, Nahar L, Shaikh K (1987) Measles case fatality among the under-fives: a multivariate analysis of risk factors in a rural area of Bangladesh. Soc Sci Med 24: 439–443.

Black FL (1966) Measles endemicity in insular populations: critical community size and its implications. J Theor Biol 11: 207–211

Black FL, Berman LL, Borgono KM et al. (1986) Geographic variation in infant loss of maternal measles antibody and in prevalence of rubella antibody. Am J Epidemiol 124: 442–452

Bottiger M, Christenson B, Romanus V, Taranger J, Strandell A (1987) Swedish experience of two dose vaccination programme aiming at eliminating measles, mumps and rubella. Br Med J 295: 1264–1267

Boue A (1964) Contribution à l'étude serologique de l' epidemiologie de la rougeole au Senegal. Bull Soc Med Afrique Noire 9: 253–254

Centers for Disease Control (1967) Recommendations of the Public Health Service Advisory Committee on Immunization Practice. Measles vaccines. Morbid Mortal Wkly Rep 16: 269–271

Chen RT, Goldbaum GM, Wassiklak SGF, Markowitz LE, Orenstein WA (1989) An explosive point-source outbreak in a highly vaccinated population: mode of transmission and risk factors for disease. Am J Epidemiol 129: 173–182

Chen RT, Weierbach R, Bisoffi Z, Cutts F, Rhodes P, Ramaroson S, Bizimana F. A (1994) "Post-honeymoon period" measles outbreak in Muyinga province, Burundi. Int J Epidemiol (23: 185-93)
Coetzee N, Yach D, Blignaut R, Fisher SA (1990) Measles vaccination coverage and its determinants in a rapidly growing peri-urban area. S Afr Med J 78: 733–737
Cutts FT, Monteiro O, Tabard P, Cliff J (1994) Measles control in Maputo, rhozambique using a single dose of Schwarz vaccine at 9 months. Bull WHO 71: 227–31
Cutts FT, Othepa O, Vernon AA et al. (1994). Measles control in kinshasa, Zaire, improved with high coverage and use of medium titer Edmonston Zagreto measles vaccine at age 6 months. Int J Epidemiol 23: 624–31
Diaz T, Nunez JC, Rullan JV, Markowitz LE, Barker ND, Horan J (1992) Risk factors associated with severe measles in Puerto Rico. Pediatr Infect Dis J 11: 836–840
Enders JF (1962) Measles virus: historical review, isolation and behavior in various systems. Am J Dis Child 103: 282–287
Enders JF, Peebles TC (1954) Propagation in tissue cultures of cytopathogenic agents from patients with measles. Proc Soc Exp Biol Med 86: 277–286
Feachem RG, Koblinsky MA (1983) Interventions for the control of diarrhoeal diseases among young children: measles immunization. Bull WHO 61: 641–652
Fine PEM (1993) Safety of measles vaccines. In: Kurstak E (ed) Measles and poliomyelitis. Springer, Vienna New York, pp 64–73
Fine PEM, Clarkson JA (1982) Measles in England and Wales-I: an analysis of factors underlying seasonal patterns. Int J Epidemiol 11: 5–14
Garenne M, Aaby P (1990) Pattern of exposure and measles mortality in Senegal. J Infect Dis 161: 1088–1094
Gordon JE, Jansen AAJ, Ascoli W (1965) Measles in rural Guatemala. Pediatrics 66: 779–786
Greenwood BM, Whittle HC (1981) Immunology of medicine in the tropics. Chapter 6: Nutrition, infection and immunity. Arnold, London, pp 179–210
Gustafson TL, Lievens AW, Bourell PA (1987) Measles outbreak in a fully immunized secondary school population. N Engl J Med 316: 717–724
Hersh BS, Markowitz LE, Maes EF, Funkhouser AW, Baughman AL, Sirotkin BI (1992) The geographic distribution of measles in the United States 1980 through 1989. JAMA 267: 1936–1941
Hope Simpson RE (1952) Infectiousness of communicable diseases in the household (Measles, chickenpox and mumps). Lancet ii: 549–554
Hull HF, Williams PJ, Oldfield F (1983) Measles mortality and vaccine efficacy in rural West Africa. Lancet 1: 972–975
Hussey GD, Klein M (1990) A randomized controlled trial of vitamin A in children with severe measles. N Engl J Med 323: 160–164
Inua M, Duggan MB, West CE et al. (1983) Post measles corneal ulceration in children in northern Nigeria: the role of vitamin A, malnutrition and measles. Ann Trop Pediatr 3: 181–191
Job JS, Halsey NA, Boulos R et al. (1991) Successful immunization of infants at 6 months of age with high dose Edmonston Zagreb measles vaccine. Pediatr Infect Dis J 10: 303–311
Kambarami RA, Nathoo KJ, Nkrumah FF, Pirie DJ (1991) Measles epidemic in spite of high measles immunization coverage rates in Harare, Zimbabwe. Bull WHO 69: 213–219
Kaschula ROC, Druker J, Kipps A (1983) Late morphological consequences of measles: a lethal and debilitating lung disease among the poor. Rev Infect Dis 5: 395–404
Khanun S, Uddin N, Garelick H et al. (1987) A comparison of Edmonston-Zagreb and Schwarz strains of measles vaccines given by aerosol or subcutaneous injection. Lancet i: 150–153
Klein-Zabban ML, Foulon G, Gôudebout C et al. (1987) Frequency of nosocomial measles in a maternal and child health center in Abidjan. Bull WHO 65: 197–201
Koster FT, Curlin GC, Aziz KMA, Haque A (1981) Synergistic impact of measles and diarrhoea on nutrition and mortality in Bangladesh. Bull WHO 59: 901–908
Krasinski K, Borkowsky W (1989) Measles and measles immunity in children infected with human immunodeficiency virus. JAMA: 261: 2512–2516
Krugman S (1983) Further attenuated measles vaccine: characteristics and use. Rev Infect Dis 5: 477–481
Krugman S, Giles JP, Jacobs AM, Friedman H (1963) Studies with a further attenuated live measles-virus vaccine. Pediatrics 31: 919–928
Krugman S, Giles JP, Friedman H, Stone S (1965) Studies on immunity to measles. J Pediatr 66: 471–488
Loutan L, Paillard S (1992) The epidemiology of measles in a nomadic community. Bull WHO 70: 741–744

Markowitz LE, Sepulveda J, Diaz-Ortega JL et al. (1990) Immunization of six-month old infants with different doses of Edmonston-Zagreb and Schwarz measles vaccines. N Engl J Med 322: 580–587
Mathias R, Meekison J, Arcan T, Schechter M (1989) The role of secondary vaccine failures in measles outbreaks. Am J Public Health 79: 475–478
McGrath D, Swanson R, Weems S, Mack D, Barbour D (1992) Analysis of a measles outbreak in Kent County, Michigan in 1990. Pediatr Infect Dis J 11: 385–389
McGregor IA (1964) Measles and child mortality in the Gambia. West Afr Med J 13: 251–257
McLean AR, Anderson RM (1988) Measles in developing countries, part II. The predicted impact of mass immunization. Epidemiol Infect 100: 419–442
Miller C (1987) Live measles vaccine: a21 years fellow-up. Br med J 295: 22–24
Mitus A, Enders JF, Craig JM, Holloway A (1959) Persistence of measles virus and depression of antibody formation in patients with giant-cell pneumonia after measles. N Engl J Med 261: 882–889.
Morley D (1973) "Severe" measles. Paediatric priorities in the developing world. Buttersworth, London.
Morley D, Woodland M, Martin WJ (1963) Measles in Nigerian children. A study of the disease in West Africa and its manifestations in England and other countries during different epochs. J Hyg (Camb) 61: 115–134
Nairn JP, Khare S. Rana SRS, Banergee KB (1989) Epidemic measles in an isolated unvaccinated population, India. Int J Epidemiol 18: 952–957
Netrasiri A, Netrasisi C (1954) Kwashiorkor in Bangkok (an analytical study of 54 cases). J Trop Pediatr 1: 148–155
Novogny T, Jennings CE, Doran M et al. (1988) Measles outbreaks in religious groups exempt from immunization laws. Public Health Rep 103: 49–54
Ogra PL, Fishaut M, Gallagher MR (1980) Viral vaccination via mucosal routes. Rev Infect Dis 2: 352–369
Panum PI (1939) Observations made during the epidemic of measles on the Faroe Islands in the year 1846. Med Class 3: 839–886
Rank EL, Brettman L, Katz-Pollack H, DeHertogh D, Neville D (1992) Chronology of a hospital-wide measles outbreak: lessons learned and shared from an extraordinary week in late March 1989. Am J Infect Control 20: 315–318
Reddy V, Bhaskaram P, Rahuramulu N et al. (1986) Relationship between measles, malnutrition, and blindness: a prospective study in Indian children. Am J Clin Nutr 44: 924–930
Reyes MA, Franky de Borrero M, Roa J, Bergonzoli G, Saravia NG (1987) Measles vaccine failure after documented seroconversion. Pediatr Infect Dis J 6: 848–851
Sabin AB, Arechiga AF, de Castro RJ et al. (1983) Successful immunization of infants with and without maternal antibody by aerosolized measles vaccine. I. Different results with undiluted human diploid cell and chick embryo fibroblast vaccines. JAMA 249: 2651–2661
Savage FMA (1967) A year of measles. Med J Zambia 1: 67–77
Scrimshaw NS, Salomon JB, Bruch HA, Gordon JE (1966) Studies of diarrhoea disease in Central America. VIII. Measles, diarrhoeal disease and nutritional deficiency in rural Guatemala. Am J Trop Med Hyg 15: 625–631
Shahid NS, Claquin P, Shaikh K, Zimicki S (1983) Long-term complication of measles in rural Bangladesh. J Trop Med Hyg 86: 77–80
Spencer HC, Kaseje DCO, Mosley WH et al. (1987) Impact on mortality and fatality of a community based malaria control programme in Saradidi, Kenya. Ann Trop Parasitol 81 [Suppl 1]: 36–45
Srivastava RN, Ghani AR, Azami S et al. (1979) Morbidity and mortality of children hospitalized with medical disorders in Afghanistan. J Trop Pediatr 25: 168–172
Sutter RW, Markowitz LE, Bennetch JM, Morris W et al. (1991) Measles among the Amish: a comparative study of measles severity in primary and secondary cases in households. J Infect Dis 163: 12–16
Taylor WR, Kalis R, ma-Disu M, Weinman JM (1988) Measles control efforts in urban Africa complicated by high incidence of measles in the first year of life. Am J Epidemiol 127: 788–794
Tidjani O, Grunitsky B, Guerin N et al. (1989) Serological effects of Edmonston-Zagreb, Schwarz, and AIK-C measles vaccine strains given at ages 4–5 or 8–10 months. Lancet ii: 1357–1360
Toole MJ, Stecketee RW, Waldman RJ, Nieburg P (1989) Measles prevention and control in emergency settings. Bull WHO 67: 381–388
Von Reyn CF, Clements CJ, Mann JM (1987) Human immunodeficiency virus infection and routine childhood immunization. Lancet ii: 669–672
Voorhoeve AM, Muller AS, Schulpen TWJ et al. (1977) Agents affecting the health of mother and child in a rural area of Kenya. III. The epidemiology of measles. Trop Geogr Med 29: 428–440
Wakeham PF (1978) Severe measles in Afghanistan. J Trop Pediatr 24: 87–88

Walsh JA (1983) Selective primary health care: strategies for control of disease in the developing world. IV. Measles. Rev Infect Dis 5: 330–340
Warner JO, Marshall WC (1976) Crippling lung disease after measles and adenovirus infection. Br J Dis Chest 70: 89–94
Whittle HC, Rowland MGM, Mann GF et al. (1984) Immunization of 4–6 month old Gambian infants with Edmonston-Zagreb measles vaccine. Lancet 2: 834–837
Whittle HC, Mann GF, Eccles M et al. (1988) Effect of dose and strain of vaccine on success of measles vaccination of infants aged 4–5 months. Lancet; 1: 963–966
WHO (1991) Measles case fatality rates. WHO, Geneva, EPI/RD/91/WP. 11
Xian J, Chen Z (1983) Measles vaccine in the People's Republic of China. Rev Infect Dis 5: 506–510
Yu-Lung Lau,Chun-Bong Chow, Ting-Hung Leung (1992) Changing epidemiology of measles in Hong Kong from 1961 to 1990 – impact of a measles vaccination programme. J Infect Dis 165: 1111–1115
Zhang Y, Su W (1983) Introduction to the control of measles by vaccination in the people's republic of China. Rev Infect Dis 5: 568–573
Zhuji Measles Vaccine Study Group (1987) Epidemiologic examination of immunity period of measles vaccine (Chinese). Chin Med J 67: 19–22

Structure, Transcription, and Replication of Measles Virus

S.M. HORIKAMI[1] and S.A. MOYER[2]

1 Virion Structure

Measles virus, a member of the morbillivirus genus of the paramyxovirus family, is an enveloped virus containing a single-stranded, minus (–) sense 50S RNA genome (BACZKO et al. 1983; DUNLAP et al. 1983; UDEM and COOK 1984). Negatively stained preparations of virus particles appear roughly spherical but pleomorphic by electron microscopy, with the diameters of the particles ranging from 300 um to 1000 um (LUND et al. 1984). The envelope of the virion consists of a lipid bilayer membrane; the integral viral membrane proteins hemagglutinin (H, 80 kDa) (GERALD et al. 1986) and the two subunits of the fusion protein (F1, 40 kDa and F2, 20 kDa) (BUCKLAND et al. 1987; RICHARDSON et al. 1986; STALLCUP et al. 1979) can be released by trypsin treatment. The matrix protein (M, 37 kDa) (GREER et al. 1986) appears to lie on the inner surface of the membrane and can be released with detergent and high salt (BELLINI et al. 1986; STALLCUP et al. 1979). Virion RNA is packaged in a helical ribonucleoprotein particle or nucleocapsid (NAKAI et al. 1969; ROBBINS et al. 1980; and LUND et al. 1984) by the nucleocapsid protein (N, 60 kDa) (ROZENBLATT et al. 1985). Measles nucleocapsids can be purified from virus banded on CsCl gradients as ribonucleoprotein particles at a density of 1.32 g/cm^3

[1] Department of Immunology and Medical Microbiology, College of Medicine, University of Florida, Gainesville, FL 32610-0266, USA
[2] Department of Pediatrics, College of Medicine, University of Florida, Gainesville, FL 32610-0266, USA

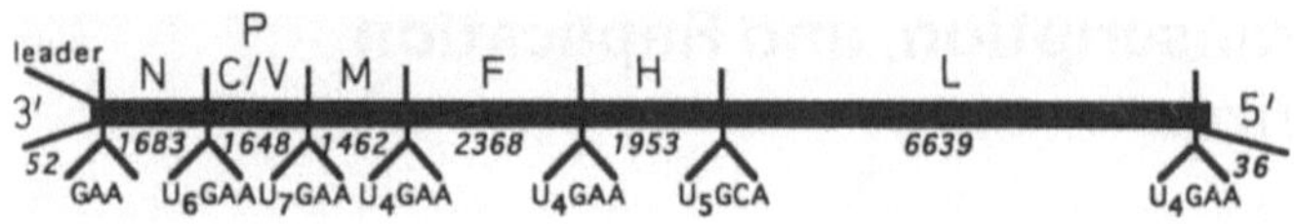

Fig. 1. Gene organization of the measles virus RNA. The measles virion RNA is indicated by the *horizontal line*, with the gene and boundary regions indicated *above* and *below the line*, respectively. The number of nucleotides in each gene is indicated *below* the gene. The data are a summary of the sequencing data contributed by the authors (referenced in the text)

(STALLCUP et al. 1979; ROBBINS et al. 1980; UDEM and COOK 1984). The association between RNA and N is very stable, being resistant to dissociation by high salt, and the encapsidated RNA is resistant to nuclease digestion (ANDZHAPARIDZE et al. 1987; MOYER et al. 1990). The virus-encoded RNA-dependent RNA polymerase consists of two subunits, the P (70 kDa) and L (~250 kDa) proteins, and is associated with the nucleocapsid in the virion (BELLINI et al. 1985; BLUMBERG et al. 1988; SEIFRIED et al. 1978).

The measles virus gene order, 3' leader-N-P/ C/ V-M-F-H-L-trailer 5' (Fig. 1), was first inferred from northern analyses of the mono- and polycistronic mRNAs isolated from measles-infected cells (BARRET and UNDERWOOD 1985; RICHARDSON et al. 1985; DOWLING et al. 1986; RIMA et al. 1986; YOSHIKAWA et al. 1986) and then directly determined by sequencing (BELLINI et al. 1985, 1986; ROZENBLATT et al. 1985; GERALD et al. 1986; ALKHATIB and BRIEDIS 1986; RICHARDSON et al. 1986;

A. Gene start Sequences

Gene	Intergenic spacer 3'	Genomic sequence 5'
N	GAA	UCCUAAGUUC.....
P/C/V	GAA	UCCUUGGUCC.....
M	GAA	UCCUCGUUUC.....
F	GAA	UCCCGGUUCC.....
H	GAA	UCCCACGUUC.....
L	GCA	UCCCAGGUUC.....

B. Gene stop Sequences

Gene	Genomic sequence 3'	5'
N	.....AGGUAGUAACAAUA	UUUUUU
P/C/V	.....GGAUUUAGGUAAUA	UUUUUUU
M	.....GGCCGUUUGAUUUG	UUUU
F	.....UAGCCAUCAAUUAA	UUUU
H	.....UGUAGUCUUAAUUC	UUUUU
L	.....UUAUAUAUAAUUUC	UUUU

Fig. 2A,B. Sequences at the 3' and 5' ends of the measles virus genes. The data from Crowley et al. (1988) are presented as (–) sense RNA

that the reaction proceeds in a standard manner, through concerted joining of the two ends of an excised linear transposon to the target DNA.

3.3 The Chemistry of DNA Breakage and Joining

The double-strand breaks that excise Tn7 from the donor backbone cleanly expose the 3' ends of the transposon but are staggered, such that 3 nucleotides of flanking donor DNA remain attached to the 5' transposon ends (BAINTON et al. 1991; P. Gary and N.L. Craig, unpublished; Fig. 4B). This contrasts with the excision of the bacterial transposon Tn10, which also transposes via an excised linear transposon where the 5' strands are also cut at the transposon termini (BENJAMIN and KLECKNER 1989, 1992). Tn7 is inserted into the target DNA by the joining of the exposed 3' transposon ends to staggered positions on the top and bottom strands of *attTn7*; these positions are separated by 5 bp (BAINTON et al. 1991). The joining of the 3' transposon ends to 5' positions in the target DNA results in the concomitant generation of exposed 3' ends in the target DNA. Thus the newly inserted Tn7 transposon is covalently linked to the target DNA through its 3' ends and is flanked by short gaps at its 5' ends, reflecting the staggered positions of end joining; the intact strands of the flanking gaps are the top and bottom strand of the target DNA between the positions of end joining. The repair of these gaps by the host DNA repair machinery results in the characteristic 5-bp duplications that flank Tn7 insertions; this repair process is also presumed to remove the few nucleotides of donor DNA attached to the 5' transposon ends in the initial recombination product.

The same chemistry of DNA breakage and joining, i.e., breakage to expose the 3' ends of the transposon and subsequent covalent linkage of these ends to 5' ends of target DNA, has been determined for all other elements that have been investigated at the biochemical level including Mu, retroviruses and retrotransposons, and other bacterial elements such as Tn10 (reviewed in MIZUUCHI 1992).

It should be noted that although it appears as if the excised transposon has joined to the target DNA by joining of the transposon ends of a staggered double-strand break at the insertion site, there is no evidence to support the view that Tn7 transposition proceeds through such a mechanism. No such double-strand

Fig. 4A, B. The pathway of transposition. **A** Transposition substrates, intermediates, and products are shown. On the *left* are the substrates, a donor plasmid containing Tn7 (white box) and a target plasmid containing *attTn7* (*black box*). Recombination initiates with double-strand breaks at either transposon end; pairs of breaks on the same substrate generate an excised linear transposon which inserts into the target DNA. The transposition products are a simple insertion and a gapped donor backbone. **B** The chemistry of breakage and joining during Tn7 transposition is shown. Recombination begins with staggered double-strand breaks at the ends of Tn7 which cleanly expose the 3' terminal-*CA*s and leave several nucleotides of donor DNA (*d*) attached to the 5' ends of the transposon. Tn7 joins to the displaced 5' positions on the target DNA though its terminal 3' *A*s. The simple insertion transposition product has covalent linkages between the 3' ends of the transposon and 5' ends of target DNA. The 5' transposon ends are flanked by short gaps; repair of these gaps by the host repair machinery generates 5-bp duplications of target sequences

the synthesis of the (+) strand nucleocapsid which in turn serves as the template for the synthesis of progeny nucleocapsids containing the (–) strand genome RNA. The newly synthesized (–) strand nucleocapsids are templates for secondary transcription, amplifying the viral mRNAs and proteins in the infected cell, as well as for budding of progeny virus.

The envelope components of the virion, the H and F proteins, are synthesized on membrane-bound ribosomes, transported through the endoplasmic reticulum, glycosylated in the Golgi, and become integral plasma membrane proteins (Morrison and Portner 1991; Ray et al. 1991b). The M protein is synthesized on cytoplasmic ribosomes and binds both progeny nucleocapsids (Stallcup et al. 1979; Hirano et al. 1992) and one or both of the viral glycoproteins at the cytoplasmic surface of the plasma membrane (Peebles 1991). The virions form by budding of the nucleocapsids through the plasma membrane, a process largely defined only by electron microscopy (Nakai et al. 1969). Virus containing both single and multiple nucleocapsids with both the (+) and (–) sense RNAs are formed, indicating a lack of specificity in the maturation process (Lund et al. 1984).

Cellular proteins are thought to be required for measles virus maturation. Cytochalasin B, an inhibitor of actin microfilaments, inhibits virus formation and results in the accumulation of measles nucleocapsids within the cell (Stallcup et al. 1983). Actin binds M protein (Guiffre et al. 1982), is also packaged in virions (Tyrrell and Norrby 1978; Stallcup et al. 1979; Robbins et al. 1980) and has been shown to be associated with measles virus nonreplicating nucleocapsids in vitro (Moyer et al. 1990). From this and electron microscopy data (Bohn et al. 1986), it is proposed that progeny nucleocapsids attach to growing actin filaments, possibly through their association with M protein bound to the nucleocapsid, and the vectoral growth of the actin filament is used as the means of transport of the nucleocapsids from the cytoplasm to the plasma membrane to initiate budding.

3 Experimental Systems for Synthesis of Viral RNA in Vitro

We will concentrate in this review specifically on the mechanisms of measles virus RNA synthesis. While many aspects of the transcription and replication of the measles virus genome appear to be similar to other nonsegmented negative strand RNA viruses, detailed studies of this virus are not as advanced. RNA synthesis of two other viruses, vesicular stomatitis virus (VSV), a rhabdovirus, and Sendai virus, a paramyxovirus, has been extensively studied (for reviews, see Banerjee 1987; Banerjee and Barik 1992; Kingsbury 1991) and these will serve as models for the present discussion when specific data on measles virus are not available. The initial studies of measles virus focused on the characterization of the various RNA products from infected cells, as summarized above, from which models for RNA synthesis were derived. Major advancements came with the establishment and characterization of a variety of cell-free systems that would support RNA synthesis in vitro. We will first discuss the various systems that have been employed and then describe the mechanisms of both transcription and RNA replication in detail.

SEIFRIED et al. (1978) first described in vitro measles transcription using detergent disrupted virus and an assay measuring the incorporation of radiolabeled GTP into TCA precipitable material. They showed that the polymerase activity copurified with virus in sucrose gradients. More recently we have also utilized purified measles virus to study various parameters of transcription in vitro (HORIKAMI and MOYER 1991). Other in vitro transcription systems utilizing cytoplasmic extracts of, or nucleocapsids from, virus-infected cells were developed and permitted the synthesis of individual virus-specific mRNAs and some genome RNA (RAY and FUJINAMI 1987; MOYER et al. 1990).

A powerful mammalian expression system utilizing cloned viral genes was developed to study VSV and Sendai virus RNA replication in vivo (PATTNAIK and WERTZ 1990; CURRAN et al. 1991). Following this methodology, we extended this system to study in vitro RNA synthesis of Sendai virus (HORIKAMI et al. 1992) and more recently of measles virus (unpublished data). In this procedure cells are infected with a recombinant vaccinia virus containing the T7 RNA polymerase gene. Newly formed T7 RNA polymerase drives expression of the subsequently transfected plasmids containing the essential Sendai virus genes NP, P, and L, which were cloned downstream of the T7 promoter. The viral proteins in extracts of these cells are used for in vitro RNA synthesis reactions. This system is very versatile, since the individual genes can be modified and then expressed in various combinations to monitor different steps in either transcription or RNA replication.

4 RNA Polymerase

The measles RNA polymerase is thought to consist of a complex formed from two different components, the L and P proteins. For VSV and Newcastle disease virus (NDV), each subunit of the RNA polymerase was purified and reconstitution experiments showed that both P and L are required for viral transcription (HAMAGUCHI et al. 1983; EMERSON and YU 1975). A similar, direct biochemical analysis of the measles virus RNA polymerase has not yet been performed. Although there is no direct experimental evidence as yet for any negative strand virus, it is believed that the L protein contains the majority, if not all, of the catalytic activities. Consistent with this hypothesis is the large size of the L protein (measles virus, 2183 amino acids). In addition, complementation studies with VSV temperature-sensitive mutants defined several intragenic complementation groups in the L gene which are also consistent with multiple functional domains in this protein (PRINGLE 1987).

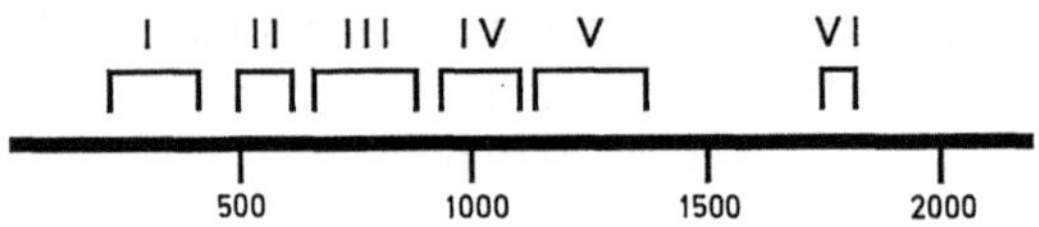

Fig. 3. Conserved regions in the measles virus L protein. The amino acid sequence of the measles L protein is depicted by the *horizontal line* with the positions of the conserved domains, I–VI, indicated *above the line*. The data were adapted from POCH et al. (1990)

The measles virus L amino acid sequence has been compared with that of other negative strand RNA viruses. Six conserved regions, I-VI (Fig. 3), have been described (BLUMBERG et al. 1988; POCH et al. 1990), although their functions have not yet been identified. The only transcription-associated activities assigned so far to the L protein have been described in VSV. The mRNA methyltransferase activities (HAMMOND and LESNAW 1987; HERCYK et al. 1988) and an aberrant polyadenylation activity (HUNT and HUTCHINSON 1993) map to the L protein. In the latter case, sequence analysis of mutants and revertants suggests that the polyadenylation mutation maps between conserved boxes V and VI in the COOH-terminus of the protein. Mutation of the GDN site in conserved region III abolished transcription of the VSV L protein (SLEAT and BANERJEE 1993), suggesting that this motif is important for catalytic activity. In other experiments CANTER et al. (1993) showed that deletion of a possible nucleotide binding motif in box VI in the VSV L protein also abrogated transcription. Future site-directed mutagenesis of the L gene in these systems will be an invaluable approach in mapping the functional domains of the protein.

Among the paramyxoviruses, the P gene is unique because it alone codes for multiple proteins, P, C and V. In measles virus the largest open reading frame in the P mRNA starts at the first AUG at nucleotide 60 and encodes the P protein (507 amino acids) in the O reading frame (BELLINI et al. 1985). In the overlapping +1 reading frame starting at nucleotide 82 is a second AUG used for the translation of the C protein (186 amino acids). The V protein will be discussed later. Both the P and C proteins are expressed from the same mRNA in measles virus-infected cells and during translation in vitro. The measles P protein is phosphorylated at multiple sites, but the C protein is unphosphorylated (BELLINI et al. 1984; ALKHATIB et al. 1988). Phosphorylation of P protein does not depend on other measles proteins, since this modification occurs when the P gene alone is expressed in an adenovirus virus vector (ALKHATIB et al. 1988). The functional role, if any, of phosphorylation is unknown, although in VSV, specific P phosphorylation sites are essential for transcription (BANERJEE and BARIK 1992).

Immunofluorescence microscopy shows that the measles C protein is present in the cytoplasm of infected cells in both diffuse and punctate staining, and it was found in the nucleus as well in one study (BELLINI et al. 1985; ALKHATIB et al. 1988). The C protein is apparently not present in purified measles virions. In Sendai virus, data from in vitro RNA synthesis reactions suggest that the C protein plays a role specifically in the regulation (inhibition) of mRNA synthesis (CURRAN et al. 1992). For measles virus we found that the addition of a rabbit anti-measles C peptide antibody to in vitro reactions had no effect on RNA synthesis (unpublished data); however, future studies using proteins expressed from cloned genes will be required to more directly address a possible role of C protein in measles virus transcription.

Although not done directly with measles virus, the P protein of the paramyxoviruses NDV and Sendai virus was shown to be required for RNA synthesis by biochemical fractionation as mentioned above and by the inhibition of transcription with monoclonal antibodies to P, respectively (HAMAGUCHI et al. 1983; DESHPANDE and PORTNER 1985). P protein appears to have multiple roles in RNA synthesis, since Sendai virus P protein forms separate complexes with the L and N proteins (HORIKAMI et al. 1992). Ryan and coworkers showed that the Sendai P protein will bind to the N protein of Sendai nucleocapsids in the absence of L protein (RYAN and PORTNER 1990; RYAN et al. 1991). Specifically, they have identified two separate regions of the P protein, one at the COOH-terminal and one at an upstream region, that are both required for binding to viral nucleocapsids. These regions are separated by a 66 amino acid domain that is not required for N binding. The individual measles virus N and P proteins also form a complex, since both full length P and a polypeptide encompassing the COOH-terminal amino acids 322–507 of P protein bound N protein in an assay in which monoclonal antibodies to either P or N coimmunoprecipitate the other protein (HUBER et al. 1991). While the N binding domain of measles P clearly resides in its COOH-terminal half, further mapping of the essential domain(s) is needed to determine if it also has the bipartite organization found in the Sendai P protein. The function of the N-P complex in RNA replication will be discussed later.

We showed for Sendai virus that RNA polymerase activity requires a complex of P and L proteins that forms only when the proteins are coexpressed in mammalian cells (HORIKAMI et al. 1992). This was demonstrated both by coimmunoprecipitation of both proteins by a monoclonal antibody to P protein and by cosedimentation of the proteins on glycerol gradients. Similar analyses of overlapping deletion mutants of the Sendai P protein expressed in mammalian cells have mapped the region between the two nucleocapsid binding regions as the L binding site on the P protein (SMALLWOOD, et al. 1994). We have recently shown for measles virus as well that the P and L proteins expressed from plasmids also form a complex (unpublished data), but it remains to be determined if the L binding site on the measles virus P protein maps to a region similar to that in Sendai virus.

5 Transcription

The measles RNA polymerase initiates transcription on the nucleocapsid RNA at the 3' terminal nucleotide, synthesizing a 55 nucleotide (+) strand leader RNA from purified virus in vitro (HORIKAMI and MOYER 1991). Free leader RNA has also been detected in measles-infected A549 cells but not in infected Vero cells (Crowley et al. 1988; Castenada and Wong 1989; Horikami and Moyer 1991), suggesting that the stability of the leader RNA in vivo is cell line-dependent. The synthesis of leader RNA demonstrates that the measles RNA polymerase initiates transcription in a manner similar to that of other non-segmented negative strand RNA viruses. Ray et al. (1991a) reported that leader RNA accumulates in the nucleus of measles virus-infected cells; however, since their assay was not specific for free leader RNA this observation requires confirmation.

Transcription then occurs sequentially following the gene order with the synthesis of the individual N, P/C/V, M, F, H, and L mRNAs. The polymerase apparently detaches from the nucleocapsid at the intergenic regions with increasing frequency as a function of the distance of the boundary from the promoter, leading to the polar accumulation of the mRNAs. mRNA abundance, therefore, is thought to be the main determinant of protein abundance. The transcription frequencies of the mRNAs in infected cells relative to N mRNA as 100%, are P (81%), M (67%), F (49%), and H (39%) (Cattaneo et al. 1987b; Schneider-Schaulies et al. 1989). A polar gradient of mRNAs was also synthesized from purified virus in vitro, except with an even steeper slope and little synthesis of the more distal mRNAs (Horikami and Moyer 1991). These results suggest that the in vitro reaction is either intrinsically inefficient or perhaps requires a cellular factor for the continued elongation or processing ability of the enzyme. At the end of each gene boundary, except for the leader RNA which is not processed, the polymerase presumably terminates synthesis following polyadenylation and then reinitiates at the consensus sequence of the next gene without transcription of the inter-genic trinucleotide (Banerjee 1987; Banerjee and Barik 1992).

As in eukaryotic mRNAs, the viral mRNAs are processed at their 5' and 3' ends. The measles virus mRNAs are capped and methylated at the 5' end with the structure 7mGpppAm (Yoshikawa et al. 1986). The mechanisms of these reactions are not known, although the conserved consensus sequence at the beginning of each gene, 3' UCC(U/C)N(G/A/C)(G/U)U(U/C)C 5' (Fig. 2A), may specify the initiation of mRNA synthesis and capping and methylation of the 5' end (Banerjee 1987). At the stop region of each gene (Fig. 2B), there is an A/U-rich sequence followed by four to seven Us, which is thought to be the signal for polyadenylation of the mRNAs (Udem and Cook 1984; Yoshikawa et al. 1986). The poly A tail varies from 70 to 140 bases (Hall and ter Meulen 1977) and appears to be synthesized by a slippage or stuttering mechanism in which the polymerase slips or backs up and copies the multiple Us reiteratively (Kolakofsky et al. 1991). It has not been determined if all these mRNA processing events occur during in vitro transcription employing purified virus, although this has been demonstrated with other negative strand viruses. Since an infectious genomic cDNA is not yet available, testing the function of the putative genomic *cis* sequences in mRNA processing has not been possible.

Measles virus transcription is not precise. Both monocistronic and various polycistronic mRNAs are synthesized in vivo and in vitro (Dowling et al. 1986; Moyer et al. 1990). Cattaneo et al. (1987a) as well as Wong and Hirano (1987) have shown for measles variant SSPE viruses that only a P-M bicistronic mRNA, and not the individual P and M mRNAs, are synthesized. Sequencing showed that the junction between the P and M regions is the exact copy of the intergenic region and polyadenylation occurs only at the end of the RNA. The synthesis of these mRNAs, therefore, is not due to mutations of the junction sequence, but seems to be a function of altered polymerases. Only the first cistron of this RNA was functional in the infected cell or in in vitro translation yielding synthesis of the P, but not the M, protein.

Transcription of the P gene of most paramyxoviruses has another unique feature, designated mRNA editing (THOMAS et al. 1988; CATTANEO et al. 1989; KOLAKOFSKY et al. 1991). CATTANEO et al. (1989) showed in measles virus-infected cells that half of the P mRNA is complementary to genomic RNA and half is altered by the insertion of a single nontemplated G nucleotide at position 751 giving a second mRNA (V) to access a third open reading frame. The putative measles virus editing sequence derived from the conserved regions in other paramyxoviruses is 5' UUAAAAAGGG*CACAGA 3' ((+) sense), in which the G insertion takes place within the three Gs indicated by the asterisk. In Sendai virus a 24 nucleotide sequence spanning the G insertion site and the three G residues were necessary and sufficient for editing in a model system (PARK and KRYSTAL 1992). We showed that P mRNA editing also occurs during in vitro transcription from purified measles virus, although the insertion of the nontemplated G occurred at a somewhat lower frequency (20%–30%) than in the infected cell (HORIKAMI and MOYER 1991). These data suggest that RNA editing is an additional function of the RNA polymerase, presumably of the L protein, as was also proposed for Sendai virus (VIDAL et al. 1990b). P mRNA editing is thought to occur by a stuttering or slippage mechanism, similar to that proposed for polyadenylation (VIDAL et al. 1990a).

The translation product of the edited mRNA, V protein shares with the P protein a common NH_2-terminal 230 amino acids but has a unique, cysteine-rich COOH-terminal polypeptide segment encompassing 68 amino acids from the third open reading frame. The V protein is phosphorylated and by immunofluorescence microscopy of measles virus-infected cells, V can be seen to have a diffuse cytoplasmic distribution distinct from that of viral nucleocapsids (WARDROP and BRIEDIS 1991; GOMBART et al. 1992). Studies on Sendai virus show that V protein is not present in virions, but is in infected cells and appears to interfere specifically with viral RNA replication. Thus, it may fulfill a regulatory role in virus reproduction (CURRAN et al. 1991); however, this remains to be determined for measles virus.

In vitro transcription of measles nucleocapsids from infected cells or of purified virus was markedly stimulated by a cytoplasmic extract of uninfected cells, suggesting the involvement of a cellular protein in viral RNA synthesis (MOYER et al. 1990; HORIKAMI and MOYER 1991; BLUMBERG et al. 1991). Tubulin appears to be at least one of the required proteins, since transcription is inhibited by an anti-tubulin monoclonal antibody and stimulated by the addition of purified tubulin. Similar results were reported for the effect of tubulin on VSV and Sendai virus RNA synthesis (MOYER et al. 1986; CHATTOPADHYAY and BANERJEE 1988; MOYER and HORIKAMI 1991). The anti-tubulin antibody coimmunoprecipitated the measles L protein with tubulin, suggesting one model in which tubulin might act as a subunit of the viral RNA polymerase (MOYER et al. 1986). Other models propose that tubulin or microtubules may act instead as an acidic activator of the RNA polymerase (CHATTOPADHYAY and BANERJEE 1988) or as an anchoring site for the transcription/replication apparatus (HAMAGUCHI et al. 1985).

6 RNA Replication

The most unusual feature of RNA replication of negative strand viruses, which distinguishes it from transcription, is that the synthesis of both the (+) and (–) genome length RNAs is coupled to their concomitant encapsidation by the nucleocapsid protein, while mRNAs remain unencapsidated. The viral polymerase responsible for genome RNA replication is thought to be the same protein complex, P-L, which catalyzes transcription and it is the availability of the N protein which is thought to regulate the transition from transcription to replication (BANERJEE 1987; BLUMBERG et al. 1981). The overall model is based on extensive data with VSV (summarized in BANERJEE 1987): The N protein, normally an insoluble protein, is complexed with P protein to form the soluble substrate used for encapsidation. To initiate encapsidation the RNA polymerase catalyzes the binding of the N protein from the complex (releasing P) to nascent leader RNA in a sequence-specific manner. The cooperative encapsidation with N coupled with further RNA synthesis masks the putative consensus signals at each gene boundary, preventing termination and mRNA processing, to yield the full-length, assembled nucleocapsid. Based on this model for RNA replication, one would predict at least three functional domains for the N protein: (1) a P binding site for the formation of the N-P complex used as the substrate for encapsidation; (2) an RNA binding site for initiation and elongation during encapsidation; and (3) an N binding (assembly) site. In addition, the RNA polymerase must recognize the N-P encapsidation substrate. None of these sites have been mapped, but some details of replication are beginning to emerge.

The measles N protein (525 amino acids) is a phosphoprotein, although the phosphorylation sites are not known. In Sendai NP, however, the nonconserved COOH-terminal region contains most of the phosphorylation sites (HSU and KINGSBURY 1982). Studies in Sendai and measles viruses suggest that the NH_2-terminal two-thirds of the N protein contain the domain that interacts with the RNA and the adjacent N polypeptides, while the nonconserved COOH-terminus (~ 100 amino acids) protrudes from, or lies on the surface of, the nucleocapsid. Trypsin treatment of the nucleocapsid, for instance, yields a large NH_2-terminal fragment still associated with the RNA and releases a COOH-terminal fragment (12K) (HEGGENESS et al. 1981; GIRAUDON et al. 1988). Deletion constructs of measles virus N have been used to map three antigenic, antibody binding sites on N protein (amino acids 122–150, 457–476, and 519–525), suggesting that there are surface epitopes at each end of the protein (BUCKLAND et al. 1989). GOMBART et al. (1993) described a different NH_2-terminal antibody binding site (amino acids 23–239) in measles N, which is conformation-dependent and recognizes N protein after it is folded and then incorporated into nucleocapsids. This is confirmation that an NH_2-terminal domain is also exposed on the surface of the nucleocapsid.

The N protein expressed from a recombinant vaccinia virus containing just the measles N gene appears to self-assemble into nucleocapsid-like structures, although it is not clear if these assembled structures contain RNA (SPEHNER et al. 1991). HUBER et al. (1991) have shown by immunofluorescence microscopy that, when measles N protein is expressed from a plasmid in transfected cells, the protein migrates predominantly to the nucleus, while the coexpression of P with N protein retains the N protein in the cytoplasm as an N-P complex. Thus, as in VSV, the P protein appears to solubilize N protein via the complex to provide a suitable substrate for replication in the proper cellular location.

It is the folded form of N protein which binds P, and the NH_2-terminus of N protein may contain the P binding site, since a monoclonal antibody to amino acids 23–239 prevents P binding (GOMBART et al. 1993). For comparison, studies in Sendai virus show that deletion of amino acids 189–293 in N protein abolished binding to the Sendai P protein, although deletions elsewhere also affected binding (HOMANN et al. 1991). Additional studies are needed to fine map the P binding site on measles N. For the other component of the complex, the COOH-terminal 322–507 amino acids of the measles P protein are sufficient to bind N protein; however, amino acids 205–507 of P are required to prevent N from migrating to the nucleus (HUBER et al. 1991). Since the V and C proteins are not synthesized from these P mutants, they appear not to be involved in N-P complex formation or its cytoplasmic retention.

The precise location of the other putative functional domains of N protein (N protein and RNA binding) have also not been determined. Sequence comparisons of the N proteins of many paramyxoviruses show a conserved amino acid sequence, spanning amino acids 261–359 (MIYAHARA et al. 1992), which may be important for these latter functions; however, no replication data are available for measles virus N protein deletions. Analyses of in vitro Sendai virus replication assays demonstrate that N deletions anywhere from amino acids 1 to 399 abolish replication, while proteins with deletions in the COOH-terminal 400–525 amino acids still give significant replication (CURRAN et al. 1993). These data support the importance of the NH_2-terminal domain in RNA binding and N-N protein interaction. The nonconserved COOH-terminus appears malleable and, in fact, can be substituted entirely by the COOH-terminus of the N protein of a bovine parainfluenza virus in vitro.

UDEM and COOK (1984) reported a ratio of two to three (–) to (+) strand nucleocapsid RNAs in measles-infected cells, suggesting that the abundance of each strand is regulated. The disproportionate synthesis of (+) and (–) nucleocapsids may simply be due to the relative level of initiation at the (+) and (–) leader templates. Although the 3' sequences of (+) and (–) RNAs are conserved for 18 nucleotides (with three changes, at positions 5,12 and 17), they diverge thereafter in the leader template (BLUMBERG et al. 1991; CROWLEY et al. 1988). The enhanced synthesis of (–) leader RNA, and thus (–) strand nucleocapsids, may occur as a result of these sequence differences.

A role for host cell proteins in RNA replication has recently been postulated. LEOPARDI et al. (1993) have shown two proteins, 22 and 30 kDa, to UV cross-link to (+) strand leader RNA, while BLUMBERG et al. (1991) found leader RNA bound to a 90 kDa cellular protein in a northwestern blot. It has been suggested that host proteins may somehow facilitate the initial binding of N protein to leader RNA to initiate replication; however, additional work is needed to establish which proteins are involved and their significance for this process.

Just as measles virus transcription seems to be inexact, since there is synthesis of polycistronic RNAs, virus infection also yields unusual replication products. In virus-infected cells, CASTENADA and WONG (1989, 1990) found novel encapsidated, but polyadenylated RNAs which contained the leader RNA fused to N RNA or to a N-P bicistronic RNA. They had a density identical to nucleocapsids, were not found on ribosomes and were apparently not translated. These RNAs seem to start as replication products but may be formed by an uncoupling of encapsidation and RNA synthesis specifically at the ends of these genes, so that the RNAs are terminated by polyadenylation.

7 Future Studies

The mammalian expression system utilizing viral genes expressed from plasmids in conjuction with biochemical assays of the parameters of RNA synthesis is a versatile, powerful approach for studying protein interactions and function. Viral genes with various mutations, deletions, truncations, or site-directed changes can be expressed and tested in vitro for transcription and replication and for binding to other proteins. The construction of an infectious measles virus cDNA is the next major research goal. This cDNA would allow the analysis of both genomic *cis* signals and the study of various gene mutations in the context of their effect on the full virus reproductive cycle. This goal should be feasible in the near future, based on the recent successes in preparing infectious subgenomic defective interfering cDNAs in several paramyxoviruses. With these reagents detailed mechanistic questions about measles virus RNA synthesis could be addressed.

Acknowledgments. We thank BRADLEY MOORE for editorial assistance and BETTINA BANKAMP for comments on the manuscript. We would also like to thank Drs. W. BELLINI, M.A. BILLETER, V. TER MEULEN, P. DOWLING and D. KOLAKOFSKY for the many valuable reagents they provided for our work. Research in the authors' laboratory was supported by grants from NIH and WHO.

References

Alkhatib G, Briedis DJ (1986) The predicted primary structure of the measles virus hemagglutinin. Virology 150: 479–490
Alkhatib G, Massie B, Briedis DJ (1988) Expression of bicistronic measles virus P/C mRNA by using hybrid adenoviruses: levels of C protein synthesized in vivo are unaffected by the presence or absence of the upstream P initiator codon. J Virol 62: 4059–4069
Andzhaparidze OG, Chaplygina NM, Bogomolova NN, Lotte VD, Koptyaeva JB, Boriskin YS (1987) Non-infectious morphologically altered nucleocapsids of measles virus from persistently infected cells. Arch Virol 95: 17–28
Baczko K, Billeter M, ter Meulen V (1983) Purification and molecular weight determination of measles virus genomic RNA. J Gen Virol 64: 1409–1413
Banerjee AK (1987) Transcription and replication of rhabdoviruses. Microbiol Rev 51: 66–87
Banerjee AK, Barik S (1992) Gene expression of vesicular stomatitis virus genome RNA. Virology 188: 417–428
Barrett T, Underwood B (1985) Comparison of meassenger RNAs induced in cells infected with each member of the morbillivirus group. Virology 145: 195–199
Bellini WJ, Englund G, Richardson CD, Rozenblatt S (1984) Positive identification of a measles virus cDNA clone encoding a region of the phosphoprotein. J Virol 50: 939–942
Bellini WJ, Englund G, Rozenblatt G, Arnheiter H, Richardson CD (1985) Measles virus P gene codes for two proteins. J Virol 53: 908–919
Bellini WJ, Englund G, Richardson CD, Rozenblatt S, Lazzarini RA (1986) Matrix genes of measles virus and canine distemper virus: cloning, nucleotide sequence, and deduced amino acid sequences. J Virol 58: 408–416
Blumberg BM, Leppert M, Kolakofsky D (1981) Interaction of VSV leader RNA and nucleocapsid protein may control VSV genome replication. Cell 23: 837–845
Blumberg BM, Crowley JC, Silverman JI, Menonna J, Cook SD, Dowling PC (1988) Measles virus L protein evidences elements of ancestral RNA polymerase. Virology 164: 487–497
Blumberg BM, Chan J, Udem SA (1991) Function of paramyxovirus 3' and 5' end sequences in theory and practice. In: Kingsbury DW (ed) The paramyxoviruses. Plenum, New York, pp 235–247
Bohn W, Rutter G, Hohenberg H, Mannweiler K, Nobis P (1986) Involvement of actin filaments in budding of measles virus: studies on cytoskeletons of infected cells. Virology 149: 91–106
Buckland R, Gerald C, Baker R, Wild TF (1987) Fusion glycoprotein of measles virus: nucleotide sequence of the gene and comparison with other paramyxoviruses. J Gen Virol 68: 1695–1703
Buckland R, Giraudon P, Wild F (1989) Expression of measles virus nucleoprotein in *Escherichia coli*: use of deletion mutants to locate the antigenic sites. J Gen Virol 70: 435–441
Canter DM, Jackson RL, Perrault J (1993) Faithful and efficient in vitro reconstitution of vesicular stomatitis virus transcription using plasmid-encoded L and P proteins. Virology 194: 518–529
Castaneda SJ, Wong TC (1989) Measles virus synthesizes both leaderless and leadercontaining polyadenylated RNAs in vivo. J Virol 63: 2977–2986
Castaneda SJ, Wong TC (1990) Leader sequence distinguishes between translatable and encapsidated measles virus RNAs. J Virol 64: 222–230
Cattaneo R, Rose JK (1993) Cell fusion by the envelope glycoproteins of persistent measles viruses which caused lethal human brain disease. J Virol 67: 1493–1502
Cattaneo R, Rebmann G, Schmid A, Baczko K, ter Meulen V, Billeter MA (1987a) Altered transcription of a defective measles virus genome derived from a diseased human brain. EMBO J 6: 681–688
Cattaneo R, Rebmann G, Baczko K, ter Meulen V, Billeter MA (1987b) Altered ratios of measles virus transcripts in diseased human brains. Virology 160: 523–526
Cattaneo R, Kaelin K, Baczko K, Billeter MA (1989) Measles virus editing provides an additional cysteine-rich protein. Cell 56: 759–764
Chattopadhyay D, Banerjee AK (1988) NH_2-terminal acidic region of the phosphoprotein of vesicular stomatitis virus can be functionally replaced by tubulin. Proc Natl Acad Sci USA. 85: 7977–7981
Crowley J, Dowling PC, Menonna J, Schanzer B, Young E, Cook SD, Blumberg BM (1987) Molecular cloning of 99% of measles virus genome, positive identification of 5' end clones, and mapping of the L gene region. Intervirology 28: 65–77
Crowley JC, Dowling PC, Menonna J, Silverman JI, Schuback D, Cook SD, Blumberg BM (1988) Sequence variablity and function of measles virus 3' and 5' ends and intercistronic regions. Virology 164: 498–506
Curran J, Boeck R, Kolakofsky D (1991) The Sendai virus P gene expresses both an essential protein and an inhibitor of RNA synthesis by shuffling modules via mRNA editing. EMBO. J 10: 3079–3085

Curran J, Marq J-B, Kolakofsky D (1992) The Sendai virus nonstructural C proteins specifically inhibit viral mRNA synthesis. Virology 189: 647–656
Curran J, Homann H, Buchholz C, Rochat S, Neubert W, Kolakofsky D (1993) The hypervariable C-terminal tail of the Sendai paramyxovirus nucleocapsid protein is required for template function but not for RNA encapsidation. J Virol 67: 4358–4364
Deshpande KL, Portner A (1985) Monoclonal antibodies to the P protein of Sendai virus define its structure and role in transcription. Virology 140: 125–134
Dörig RE, Marcil A, Chopra A, Richardson CD (1993) The human CD 46 molecule is a receptor for measles virus (Edmonston strain). Cell 75: 295–305
Dowling PC, Blumberg BM, Menonna J, Adamus JE, Cook PJ, Crowley JC, Kolakofsky D, Cook SD (1986) Transcriptional map of the measles virus genome. J Gen Virol 67: 1987–1992
Dunlap RC, Milstien JB, Lundquist ML (1983) Identification and characterization of measles virus 50S RNA. Intervirology 19: 169–175
Emerson SU, Yu YH (1975) Both NS and L proteins are required for in vitro RNA synthesis by vesicular stomatitis virus. J Virol 15: 1348–1356
Galinski M (1991) Annotated nucleotide and protein sequneces for selected paramyxoviridae. In: Kingsbury DW (ed) The paramyxoviruses. Plenum, New York, pp 537–568
Gerald C, Buckland R, Barker R, Freeman G, Wild TF (1986) Measles virus haemagglutinin gene: cloning, complete nucleotide sequence analysis and expression in Cos cells. J Gen Virol 67: 2695–2703
Giraudon P, Jacquier MF, Wild TF (1988) Antigenic analysis of African measles virus field isolates: identification and localisation of one conserved and two variable epitope sites on the NP protein. Virus Res 18: 137–152
Gombart AF, Hirano A, Wong TC (1992) Expression and properties of the V protein in acute measles virus and subacute sclerosing panencephalitis virus strains. Virus Res 25: 63–78
Gombart AF, Hirano A, Wong TC (1993) Conformational maturation of measles virus nucleocapsid protein. J Virol 67: 4133–4141
Greer PA, Hasel KW, Millward S (1986) Cloning and in vitro expression of the measles virus matrix gene. Biochem Cell Biol 65: 1038–1043
Guiffre RA, Tovell DR, Kay CM, Tyrrell DLJ (1982) Evidence for an interaction between the membrane protein of a paramyxovirus and actin. J Virol 42: 963–968
Hall WW, ter Meulen V (1977) Polyadenylic acid [poly (A)] sequences associated with measles virus intracellular ribonucleic acid (RNA) species. J Gen Virol 35: 497–510
Hamaguchi M, Yoshida T, Nishikawa K, Naruse H, Nagai Y (1983) Transcriptive complex of Newcastle disease virus I. Both L and P proteins are required to reconstitute an active complex. Virology 128: 105–117
Hamaguchi M, Nishikawa K, Toyoda T, Yoshida T, Hanaichi T, Nagai Y (1985) Transcriptive complex of Newcastle disease virus II. Structural and functional assembly associated with the cytoskeletal framework. Virology 147: 295–308
Hammond OC, Lesnaw JA (1987) Functional analysis of hypomethylation variants of the New Jersey serotype of vesicular stomatitis virus. Virology 160: 330–335
Hasel KW, Day S, Millward S, Richardson CD, Bellini WJ, Greer PA (1987) Characterization of cloned measles virus mRNAs by in vitro transcription, translation and immuno-precipitation. Intervirology 28: 26–39
Heggeness MH, Scheid A, Choppin PW (1981) The relationship of conformational changes in the Sendai virus nucleocapsid to proteolytic cleavage in the NP polypeptide. Virology 114: 555–562
Hercyk N, Horikami SM, Moyer SA (1988) The vesicular stomatitis virus L protein possesses the mRNA methyltransferase activities. Virology 163: 222–225
Hirano A, Wang AH, Gombart AF, Wong TC (1992) The matrix proteins of neurovirulent subacute sclerosing panencephalitis virus and its acute measles virus progenitor are functionally different. Proc Natl Acad Sci USA 89: 8745–8749
Homann HE, Willenbrink W, Buchholz CJ, Neubert WJ (1991) Sendai virus protein–protein intractions studied by a protein-blotting protein-overlay technique: mapping of domains on NP protein required for binding to P protein. J Virol 65: 1304–1309
Horikami SM, Moyer SA (1991) Synthesis of leader RNA and editing of the P mRNA during transcription by purified measles virus. J Virol 65: 5342–5347
Horikami SM, Curran J, Kolakofsky D, Moyer SA (1992) Complexes of Sendai virus NP-P and P-L proteins are required for defective interfering particle genome replication in vitro. J Virol 66: 4901–4908
Hsu, C-H, Kingsbury DW (1982) Topography of phosphate residues in Sendai viral proteins. Virology 120: 225–234

Huber M, Cattaneo R, Spielhofer P, Örvell C, Norrby E, Messerli M, Perriard J-C, Billeter MA (1991) Measles virus phosphoprotein retains the nucleocapsid protein in the cytoplasm. Virology 185: 299–308
Hunt DM, Hutchinson KL (1993) Amino acid changes in the L polymerase proteins of vesicular stomatitis virus which confer aberrant polyadenylation and temprature-sensitive phenotypes. Virology 193: 786–793
Kingsbury DW (ed) (1991) The paramyxoviruses. Plenum, New York
Kolakofsky D, Vidal S, Curran J (1991) Paramyxovirus RNA synthesis and P gene expression. In: Kingsbury DW (ed) The paramyxoviruses. Plenum, New York, pp 215–233
Leopardi R, Hukkanen V, Vaenconpää R, Salmi AA (1993) Cell proteins bind to sites within the 3' noncoding region and the positive-strand leader sequence of measles virus RNA. J Virol 67: 785–790
Lund GA, Tyrrell DLJ, Bradley RD, Scraba DG (1984) The molecular length of measles virus RNA and the structural organization of measles nucleocapsids. J Gen Virol 65: 1535–1542
Miyahara K, Kitada S, Yoshimoto M, Matsumura H, Kawano M, Komada H, Tsurudome M, Kusugawa S, Nishio M, Ito Y (1992) Molecular evolution of paramyxoviruses: nucleotide sequence analyses of the human parainfluenza type I virus NP and M protein genes and construction of phylogenetic trees for all the human paramyxoviruses. Arch Virol 124: 255–268
Morrison T, Portner A (1991) Structure, function and intracellular processing of the glycoproteins of paramyxoviridae. In: Kingsbury DW (ed) The paramyxoviruses. Plenum, New York, pp 347–382
Moyer SA, Horikami SM (1991) The role of viral and host cell proteins in paramyxovirus transcription and replication. In: Kingsbury DW (ed) The paramyxoviruses. Plenum, New York, pp 249–274
Moyer SA, Baker SC, Lessard JL (1986) Tubulin: a factor necessary for the synthesis of both Sendai virus and vesicular stomatitis virus RNAs. Proc Natl Acad Sci USA 83: 5405–5409
Moyer SA, Baker SC, Horikami SM (1990) Host cell proteins required for measles virus reproduction. J Gen Virol 71: 775–783
Nakai T, Shand FL, Howatson AF (1969) Development of measles virus in vitro. Virology 38: 50–67
Naniche D, Varior–Krishnan G, Cervoni F, Wild TF, Ross B, Rabourdin–Combe C, Gerlier D (1993) Human membrane cofactor protein (CD 46) acts as a cellular receptor for measles virus. J Virol 67: 6025–6032
Park KH, Krystal M (1992) In vivo model for pseudo-templated transcription in Sendai virus. J Virol 66: 7033–7039
Pattnaik AK, Wertz GW (1990) Replication and amplification of defective interfering particle RNAs of vesicular stomatitis virus in cells expressing viral proteins from vectors containing cloned cDNAs. J Virol 64: 2948–2957
Peeples ME (1991) Paramyxovirus M proteins: Pulling it all together and taking it on the road. In: Kingsbury DW (ed) The paramyxoviruses. Plenum, New York, pp 427–456
Poch O, Blumberg BM, Bougueleret L, Tordo N (1990) Sequence comparison of five polymerases (L proteins) of unsegmented negative-strand RNA viruses: theoretical assignment of functional domains. J Gen Virol 71: 1153–1162
Pringle CR (1987) Rhabdovirus genetics. In: Wagner RR (ed) The rhabdoviruses. Plenum, New York, pp 167–243
Ray J, Fujinami RS (1987) Characterization of in vitro transcription and transcriptional products of measles virus. J Virol 61: 3381–3387
Ray J, Whitton JL, Fujinami RS (1991a) Rapid accumulation of measles leader RNA in the nucleus of infected HeLa cells and human lymphoid cells. J Virol 65: 7041–7045
Ray R, Roux L, Compans RW (1991b) Intracellular targeting and assembly of paramyxovirus proteins. In: Kingsbury DW (ed) The paramyxoviruses. Plenum, New York, pp 457–480
Richardson CD, Berkovich A, Rozenblatt S, Bellini WJ (1985) Use of antibodies directed against synthetic peptides for identifying cDNA clones, establishing reading frames and deducing the gene order of measles virus. J Virol 54: 186–193
Richardson C, Hall D, Greer P, Hasel K, Berkovich A, Englund G, Bellini W, Rima B, Lazzarini R (1986) The nucleotide sequence of the mRNA encoding the fusion protein of measles virus (Edmonston strain): a comparison of fusion proteins from several different parmyxoviruses. Virology 155: 508–523
Rima BK, Baczko K, Clarke DK, Curran MD, Marten SJ, Billeter MA, ter Meulen V (1986) Characterization of clones for the sixth (L) gene and a transcriptional map for morbilliviruses. J Gen Virol 67: 1971–1978
Robbins SJ, Bussell RH, Rapp F (1980) Isolation and partial characterization of two forms of cytoplasmic nucleocapsids from measles virus-infected cells. J Gen Virol 47: 301–310

Rozenblatt S, Eizenberg O, Ben-Levy R, Lavie V, Bellini WJ (1985) Sequence homology with the morbilliviruses. J Virol 53: 684–690
Ryan KW, Portner A (1990) Separate domains of Sendai virus P protein are required for binding to viral nucleocapsids. Virology 174: 515–521
Ryan KW, Morgan EM, Portner A (1991) Two noncontiguous regions of Sendai virus P protein combine to form a single nucleocapsid binding domain. Virology 180: 126–134
Schneider-Schaulies S, Liebert UG, Baczko K, Cattaneo R, Billeter M, ter Meulen V (1989) Restriction of measles virus gene expression in acute and subacute encephalitis of Lewis rats. Virology 171: 525–534
Seifried AS, Albrecht P, Milstien JB (1978) Characterization of an RNA-dependent RNA polymerase activity associated with measles virus. J Virol 25: 781–787
Sleat DE, Banerjee AK (1993) Transcriptional activity and mutational analysis of recombinant vesicular stomatitis virus RNA polymerase. J Virol 67: 1334–1339
Smallwood S, Ryan KW, Mayer SA (1994) Deletion analyses defines a carboxyl, proximal region of sendai virus P protein that binds to the polymerase L protein. Virology 202: 154–163
Spehner D, Kirn A, Drillien R (1991) Assembly of nucleocapsidlike structures in animal cells infected with a vaccinia virus recombinant encoding the measles virus nucleoprotein. J Virol 65: 6296–6300
Stallcup KC, Wechsler SL, Fields BN (1979) Purification of measles virus and characterization of subviral components. J Virol 30: 166–176
Stallcup KC, Raine CS, Fields BN (1983) Cytochalasin B inhibits the maturation of measles virus. Virology 124: 59–74
Thomas SM, Lamb RA, Paterson RG (1988) Two mRNAs that differ by two nontemplated nucleotides encode the amino coterminal proteins P and V of the paramyxovirus SV5. Cell 54: 891–902
Tyrrell DLJ, Norrby E (1978) Structural polypeptides of measles virus. J Gen Virol 39: 219–229
Udem SA, Cook KA (1984) Isolation and characterization of measles virus intracellular nucleocapsid RNA. J. Virol 49: 57–65
Vidal S, Curran J, Kolakofsky D (1990a) A stuttering model for paramyxovirus P mRNA editing. EMBO J 9: 2017–2022
Vidal S, Curran J, Kolakofsky D (1990b) Editing of the Sendai virus P/C mRNA by G insertion occcurs during mRNA synthesis via a virus encoded activity. J Virol 64: 239–246
Wardrop EA, Briedis DJ (1991) Characterization of V protein in measles virus-infected cells. J Virol 65: 3421–3428
Wild TF, Malvoisin E, Buckland R (1991) Measles virus: both the haemagglutinin and the fusion glycoprotein are required for fusion. J Gen Virol 72: 439–442
Wong TC, Hirano A (1987) Structure and function of bicistronic RNA encoding the phosphoprotein and matrix protein of measles virus. J Virol 61: 584–589
Yoshikawa Y, Mizumoto K, Yamanouchi K (1986) Characterization of messenger RNAs of measles virus. J Gen Virol 67: 2807–2812

Functional Aspects of Envelope-Associated Measles Virus Proteins

T.F. WILD and R. BUCKLAND

1 Introduction

Measles virus (MV) exhibits a very limited host range. Humans are the only known reservoir, although the virus can infect some monkey species. In the laboratory, the virus produces characteristic multinucleated giant cells and intracellular inclusion bodies when grown in a variety of human and simian cell cultures, thereby confirming this host specificity. The cytopathic affects are the result of a chain of events. Attachment of the virus is followed by a replicative cycle and the release of virus particles at the cell membrane. These steps all involve specific interaction of virus proteins, either between themselves or with cellular proteins. In this review we shall emphasize the importance of these protein–protein interactions for the function of the envelope-associated MV proteins. The inability of certain virus isolates, such as strains from subacute sclerosing panencephalitis (SSPE) patients, to sustain one or more of these interactions leads to incomplete or defective persistent infections in which virus particles are not formed.

MV is a member of the morbillivirus subgroup of the paramyxovirus family, distinguished from the major group by a lack of neuraminidase activity. Other members include rinderpest (RPV), canine distemper virus (CDV), peste des

Inserm U404, Immunité et Vaccination, Institut Pasteur, Avenue Tony Garnier, 69365 Lyon Cedex 07, France petits ruminants and the recently isolated phocine distemper virus (PDV) and the dolphin morbillivirus (DMV). They constitute a family of immunologically related viruses, with the antigenic differences between members used to determine their evolution. Thus it has been proposed that RPV is the archevirus of the morbilliviruses from which CDV was the first to adapt to a new host, to be followed later by MV (NORRBY et al. 1985). Passage into a new host would be dependent on changes in certain viral proteins especially those responsible for direct virus-host cell interaction.

It is useful, when examining the biological function of MV proteins, to compare not only the divergence between the closely related morbilliviruses but also the more distantly related paramyxoviruses. Comparison of the primary sequences for a given protein with those of other members of the virus family can aid in the identification of elements such as glycosylation sites, hydrophobic sequences and certain amino acids (cysteine, proline and glycine) which are important structural determinants. The actual demonstration of functional domains on proteins and the precise identification of the amino acids involved have been greatly aided by the mapping of epitopes for neutralizing monoclonal antibodies (MAbs) and the sequencing of mutants defective for specific functions.

2 The Hemagglutinin Protein

Although the primary function of the MV hemagglutinin (H) protein is attachment to the MV receptor on the host cell, in terms of viral replication its interaction with the MV fusion protein is of equal functional importance.

The MV H glycoprotein is a type 2 membrane protein of 617 amino acids in which the five potential glycosylation sites are grouped between amino acids 168 and 238; the last is not actually used. After synthesis and glycosylation in the endoplasmic reticulum disulphide-linked dimers are formed and are transported to the Golgi for oligosaccharide modification. The H protein is then transported to the cell membrane, where it can be isolated as tetramer (OGURA et al. 1991).

Studies on wild-type strains isolated in 1983 and 1984 in Africa failed to show any antigenic variation of the H protein compared with the vaccine strain, confirming the antigenic stability of this glycoprotein (GIRAUDON et al. 1988). However, when the rate of nucleotide change was measured for measles viruses circulating in the USA, it was found that the rate had doubled over the last 40 years (ROTA et al. 1992). The observed mutation rate of 0.13% is within an order of magnitude of the rates calculated for the H protein of influenza A virus, which is known to respond to immunological pressure (WILSON and COX 1990). The majority of amino acid changes were clustered near the region encoding the potential glycosylation sites. The nonrandom distribution of these conserved coding changes may be indicative of selective pressure.

Although only minor changes in the H protein are observed between different MV isolates, amongst the different members of the morbillivirus group the H protein displays the greatest diversity. The majority of MAbs against the H proteins of MV, RPV or CDV do not cross-react with other members of the group (SHESHBERADARAN et al. 1985). Sequence studies have shown that the MV H protein has 60% homology with RPV and 37% with CDV (GERALD et al. 1986; TSUKIYAMA et al. 1987; KÖVAMEES et al. 1991 a,b). This variability in the protein responsible for cell attachment probably reflects the specificity of the virus for different hosts and thus for different receptors. Normally, when different proteins possess the same function, there is a corresponding conservation of functional domains between them. In the H protein the lack of immunological cross-reactivity is mirrored by a dearth of conserved elements, making the assignment of functional domains problematic. However, a single hydrophobic domain close to the NH_2-terminal representing the transmembrane region is present in the MV H protein and is conserved in all paramyxovirus attachment proteins. There is no signal peptide at the NH_2-terminal, but presumably the transmembrane region fulfils this function as in other type 2 membrane proteins (ZERIAL et al. 1987). Also the conservation of 12 cysteines and approximately half of the glycine and proline residues between the morbillivirus H proteins suggests that at least the proteins have a similar tertiary structure. The glycosylation sites in the MV H protein are all within a region of 70 amino acids, a short distance COOH-terminal from the transmembrane region. Bunching of the glycosylation sites to a small region in the extracellular domain of the H protein is typical of the morbillivirus H protein. This is in contrast to the other paramyxovirus attachment proteins, in which the sites are more randomly distributed, at least as far as the primary sequence is concerned. Presumably this relates to a structure and function particular to the morbillivirus H protein, but whether this involves the lack of neuraminidase activity is unknown.

Information on the location of functional domains in the MV H protein can be obtained by epitope mapping with anti-H protein MAbs. Until recently, it was assumed that neutralizing anti-H protein MAbs block attachment to the cellular receptor. It has now been established that the interaction of H protein with fusion (F) protein is necessary for fusion to take place (WILD et al. 1991). The abrogation of this supposed interaction should also be considered as a target for these MAbs. The MV H protein is capable of agglutinating red blood cells (RBCs) from a variety of monkeys (e.g., vervet, patas) but not human or other species. This is because the MV receptor (see below) is present on monkey RBCs but absent on human RBCs. MAbs which block hemagglutination or hemadsorption and which also neutralize virus infectivity either in vitro or in vivo are directed against conformational epitopes (GIRAUDON and WILD 1985).

To study the regions implicated in the interaction between the H protein and its receptor, a number of epitopes have been determined on the H protein (GIRAUDON and WILD 1985 ; SHESHBERADARAN and NORRBY 1986). However, it is not yet clear if the different epitopes identified represent distinct separate entities or are all part of the same conformational site. Sequential removal of eight epitopes did not affect hemagglutination properties or virus growth in culture, indicating that changes in these epitopes do not affect the binding site. As polyclonal serum recognized even the most mutated H protein, it was concluded that the monotypic status of MV is not disturbed by the existence of variants with rather extensive epitope changes in their H protein (SHESHBERADARAN and NORRBY 1986). Despite these observations, hemagglutination activity can be distinguished from certain immunological reactivity. Incubation of the H at pH 4.5 at 37°C causes irreversible antigenic changes and the loss of hemagglutination activity. The induction of the antigenic changes lags behind the loss of hemagglutination activity and does not affect a number of the epitopes (SHESHBERADARAN 1991). Similarly, SUGIYAMA et al. (1991) have shown that after treatment of RPV H protein with mercaptoethanol, some epitopes are destroyed but not others. These observations indicate that certain epitopes on the HA are more resistant to alteration that others.

Attempts to localize epitopes on the MV H protein have met with little success due to the conformational nature of the antigen. However, MÄKELÄ et al. (1989) mapped the reactivity of three MAbs using synthetic peptides to regions 126–135, 309–318 and 587–596. HU et al. (1993), sequencing escape mutants, also found mutations within the 309-318 epitope at 313–314 and others at 189 and 491–552. It is not yet known if any of these sites relate to functional domains.

2.1 Characterization of the Measles Virus Receptor

Early studies to identify the MV receptor used an antiidiotype approach. Two polypeptides of 20 and 30.5 kDa were immunoprecipitated by this technique, but not identified (KRAH and CHOPPIN 1988). More recently, MAbs were prepared against Jurkat cells (human lymphoblasts) and screened for their ability to block fusion induced in HeLa cells infected with a vaccinia recombinant expressing MV H and F proteins (NANICHE et al. 1992). MAbs which inhibited this activity were shown to recognize the human membrane cofactor protein CD46 (NANICHE et al. 1993b). CD46, which has a molecular mass of 57/67 kDa is a widely distributed regulatory glycoprotein of the human complement system. The biological role of CD46 (Fig. 1) is to prevent complement-mediated lysis by acting as a cofactor for plasma serine protease factor I (for a review see LISZEWSKI et al. 1991). CD46 is a type 1 membrane glycoprotein in which the NH_2-terminal two-thirds are composed of four short consensus repeats (SCRs). Each SCR is a cysteine-rich repeating domain of approximately 60 amino acids. The four SCRs precede a serine/threonine/proline-rich domain, a region of potential heavy O-glycosylation. Variability in this domain and the cytoplasmic tail is generated by alternative splicing, which gives rise to multiple protein isoforms. Peripheral blood cells regularly express variable quantities of four primary isoforms. The significance, if any, of the different isoforms in terms of the MV receptor has not yet been investigated.

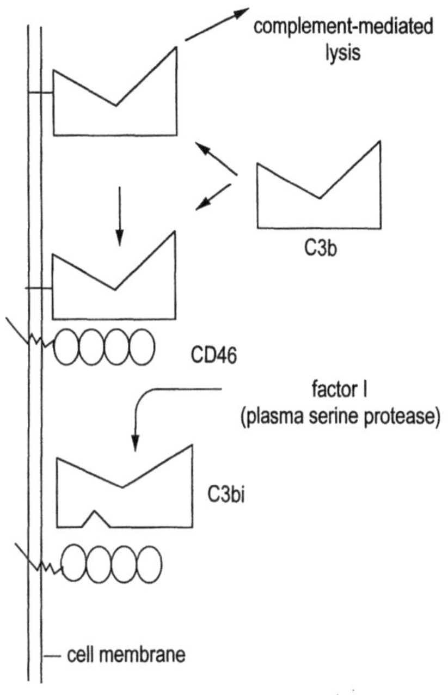

Fig. 1. The action of CD46 in its abrogation of complement-mediated lysis. The *C3b* protein, a component of the complement cascade, forms a complex with *CD46* after attaching to the cell membrane. Once associated with CD46, C3b becomes susceptible to cleavage by factor 1 and is thus inactivated (C3bi)

The specificity of CD46 as a receptor for MV was confirmed by establishing murine cell lines constitutively expressing CD46 protein. MV attaches specifically to these cells but not to control nontransfected cells. With the exception of L cells, murine CD46$^+$ cells are permissive for MV replication. As infection of murine CD46$^+$ (P815, M12, 3T3, L) cells with a vaccinia recombinant encoding both H and F proteins leads to syncytia formation (NANICHE et al. 1993b), the block of MV infection in L cells is probably not at the entry level but at a later stage in the virus cycle.

2.2 Functional Consequences of H Protein and Measles Virus Receptor Interaction

The continuing presence of a virus receptor at the cell surface may determine the fate of the viral infection. Thus, it has been suggested that removal of the human parainfluenza virus 3 receptor during infection can prevent cell-cell fusion and lead to a persistent state (MOSCONA and PELUSO 1991). This phenomenon, whereby limitation of the receptor at the cell surface results in the establishment of persistence, has also been observed with HIV infections: the CD4 receptor for HIV is down-regulated from the cell surface by several mechanisms (HOXIE et al. 1986; CRISE et al. 1990; GARCIA and MILLER 1991). In cell lines persistently infected with MV, cell-cell fusion did not occur unless fresh cells were added to the culture (HIRANO 1992) and recently it was shown that infection of HeLa or Jurkat cells with MV leads to the down-regulation of CD46 molecules expressed at the cell surface

(NANICHE et al. 1993a). This effect was specific for MV as it was not seen with cells infected with CDV. The decrease in cell surface CD46 correlated with the expression of MV glycoproteins: expression of H protein from a vaccinia virus recombinant was sufficient to induce the down-regulation. Cell surface expression kinetics of H protein and CD46 showed that the degree of down-regulation was correlated with the amount of H protein expressed in the infected cells. During infection the down-regulation occurred by rapid internalization of CD46 from the cell surface (NANICHE et al. 1993a). The removal of the MV receptor (CD46) from the cell surface may have two opposing effects: although infected cells with decreased CD46 will be more susceptible to complement-mediated cell lysis, such cells will also have reduced fusability, which could favor establishment of viral persistence.

3 The Fusion Protein

The F glycoprotein is a type 1 membrane protein responsible for fusion of the virion and host cell membranes. This process occurs at neutral pH and results in the liberation of the viral capsid into the cytoplasm of the host cell.

After synthesis and glycosylation in the endoplasmic reticulum as inactive F_0 precursor, the protein is transported to the Golgi where cellular subtilisin-related proteases cleave F_0 into two disulphide-linked subunits, F_1 and F_2. All of the potential N-glycosylation sites are contained in the F_2 subunit. The fully processed $F_{1,2}$ protein arrives at the cell membrane as an oligomer, a tetramer possibly a trimer (SECHOY et al. 1987; BUCKLAND et al. 1992). A number of the F proteins of the cell surface are uncleaved, but despite this, virions do not normally contain uncleaved F_0 (GRAVES et al. 1978). This would seem to suggest a positive selection of cleaved F proteins for incorporation into the virion. In contrast, MV infection of lymphoblast cell lines which lack the appropriate cellular protease to cleave F_0 leads to an infection with the absence of syncytia, but with the release of virus particles containing F_0 (FUJINAMI and OLDSTONE 1981). Thus, lack of cleavage does not inhibit the incorporation of F into virus particles. The virus particles from such cells are noninfectious but can become so if incubated with trypsin, which correctly cleaves the virion F protein.

The functional activity of the F protein imposes certain structural restraints on the protein. Such is the restriction that within the morbillivirus group there is at least 70% conservation of the amino acids (BARRETT et al. 1987; BUCKLAND et al. 1987; TSUKIYAMA et al. 1988; KÖVAMEES et al. 1991a). Comparison of F protein primary sequences from different paramyxoviruses reveals a number of conserved structurally important elements which aid in the assignment of domains (MORRISON and PORTNER 1991). Presumably each domain in the MV F protein (Fig. 2) reflects a biological function and whereas the function of some are known, the function of others is open to speculation. There are three hydrophobic regions: (1) a signal peptide at the NH_2-terminal of the F_2 subunit, which directs membrane

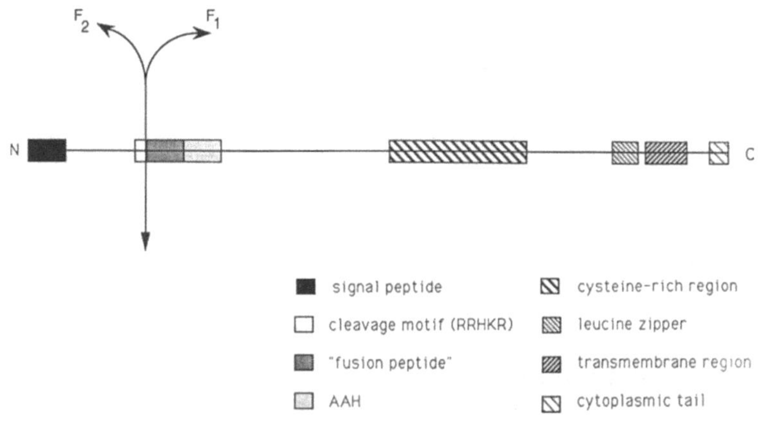

Fig. 2. Linear representation of the MV F_0 showing structural elements which may be functionally important. Note that all the potential N-glycosylation sites are in the F_2 subunit

insertion; (2) the "fusion peptide" at the NH_2-terminal of the F_1 subunit, which is implicated in the fusion process; and (3) a transmembrane region at the COOH-terminal of F_1, responsible for anchoring the protein in the viral and plasma membranes. There are other domains for which the function is not so clear, including the last 14 amino acids in the cytoplasmic tail. These are conserved in all morbillivirus F proteins and are possibly responsible for interaction with the matrix protein, although as yet no such association has been demonstrated. In addition, there is a cysteine-rich region of eight cysteines between positions 337 and 423. Preliminary results from a study on the interaction of F protein with H protein (BUCKLAND et al., unpublished data) suggest that the cysteine-rich region plays a role in this association. If true, this could explain why escape mutants to fusion-inhibiting MAbs directed against the Sendai virus F protein (PORTNER et al. 1987) and the NDV F protein (TOYODA et al. 1988; NEYT et al. 1989) map in this region. It can also be speculated that such closely situated cysteines could provide a degree of local flexibility in the protein, perhaps useful when traversing membranes, by a rapid making and breaking of disulfide bonds between the different cysteines, as in the proposed "molten globule" state of protein structure (EWBANK and CREIGHTON 1991). There are also two amphipathic α-helices present in the MV F protein, one of which (AAH) is COOH-terminal to the fusion peptide and the other NH_2-terminal to the transmembrane region. Due to the presence of heptadic leucines, the latter is known as the "leucine zipper" (Buckland and WILD 1989). This structure was originally suggested to play a role in the oligomerization of the F protein (BUCKLAND and WILD 1989) but has now been shown to be essential for the process of fusion (BUCKLAND et al. 1992). As AAH is COOH-terminal to the fusion peptide, this could also play a role in the mechanism of fusion.

Another region which is well conserved is that around the cleavage site. The sequence preceding the cleavage site, i.e., the COOH-terminal of the F_2 subunit, is important for the specificity of the cleavage. In the MV F protein, the sequence is RRHKR. GOTOH et al. (1992) have shown that such oligobasic cleavage motifs are best cleaved by the subtilisin-related protease Furin/PACE. The 25 hydrophobic amino acids that follow the cleavage site and which become the NH_2-terminal of the F_1 subunit following cleavage represent the fusion peptide, presumed to be primarily responsible for the subsequent fusion activity. HSU et al. (1981) showed that when the Sendai F_0 is cleaved to $F_{1,2}$ there is a concomitant conformational change and increase in α-helical content which is due to the liberation of the fusion peptide. It is thought that the fusion peptide can then interact directly with the target membrane inducing fusion (GETHING et al. 1978).

4 Measles Virus Fusion

Viral fusion is not the function of a sole MV protein, as expression of the MV F protein alone from a vaccinia recombinant does not lead to cell fusion. Fusion can only be produced by recombinants expressing both H and F proteins (TAYLOR et al. 1991; WILD et al. 1991). Furthermore, as fusion does not occur if the cells used do not express the MV cellular receptor, it can be assumed that a ternary complex consisting of H, F and receptor is a prerequisite for the process. That is not to say that the presence of all three proteins is strictly essential, as it has been reported that expression of the F protein alone in certain circumstances can lead to fusion. High expression of the MV F protein via an adenovirus vector induced fusion (ALKHATIB et al. 1990). Furthermore, although infection of insect cells (which presumably lack the MV receptor), with a baculovirus expressing MV F protein did not trigger fusion, if the pH of the infected cultures was lowered to 5.8, fusion was observed (VIALARD et al. 1990). This may indicate that the H protein and receptor fulfil a positioning role to optimize the conditions for fusion, but they can be dispensed with if the F protein is present in high enough numbers or is suitably orientated in the membrane.

Thus, the function of H protein in the fusion process may be simply to anchor the infected cell to the receptor of the target cell and then associate with the F protein in such a way that the latter is placed in a position where it can interact with the cell membrane. Studies with human parainfluenza 2 and 3 viruses have shown that fusion results when the HN and F proteins of each virus are expressed in the homologous system, but not when the F protein is expressed with the HN protein of the other serotype (HU et al. 1992). Our own experiments with MV and CDV have shown that the coexpression of MV H and CDV F proteins does not lead to fusion. However, when regions of the CDV F gene are exchanged with their homologues in MV and subsequently coexpressed with MV H protein, fusion takes place (BUCKLAND et al., unpublished results). In addition, when H and F variants derived from cases of subacute sclerosing panencephalitis (SSPE) were coexpressed, only certain combinations of H and F variants gave rise to fusion whereas others did not (CATTANEO and ROSE 1993). All these studies suggest that the two MV glycoproteins interact specifically with each other at the cell membrane. CITOVSKY et al. (1986) used circular dichroism to demonstrate that when Sendai HN and F proteins associate in the same membrane, conformational changes are induced in both glycoproteins. Studies with both measles and Sendai viruses have shown that the H and F proteins processed in the cell at different rates, the F protein arriving at the surface faster that the H protein (MOTTET et al. 1986; SATO et al. 1988).

This would suggest that if the H and F proteins interact, it must be a late event and does not occur during transport. "Fusion from without", i.e., virus-cell fusion, appears to differ from "fusion from within," or cell-cell fusion (MOSCONA and PELUSO 1991). Lectins are effective against cell-to-cell fusion but not against virus-cell fusion. If there is a need for the H protein to associate with the F protein in the membrane, this could be blocked by lectins which prevent the lateral movement required for the two glycoproteins to associate. In the virion, however, the two glycoproteins are already associated and therefore can function immediately in attachment and fusion.

Once in a favorable context for fusion, how does the F protein interact with the cell membrane? Since the hydrophobic region at the NH_2-terminal region of F_1 exposed after cleavage is highly conserved amongst paramyxovirus F proteins, it was suggested that this region (the fusion peptide) could promote fusion with target membranes (GETHING et al. 1978). Subsequently, RICHARDSON et al. (1980) showed that synthetic peptide analogues of the fusion peptide sequence inhibit both cell fusion and virus penetration but not virus attachment. Furthermore, NOVICK and HOEKSTRA (1988), using photoactive lipid probes, showed that the fusion peptide interacted with target lipid bilayers. In an effort to understand the mechanism of fusion, HULL et al. (1987) isolated a MV mutant which was resistant to inhibition by the oligopeptide analogue. However, sequencing of the F gene from this mutant revealed the presence of three amino acid changes, none of which were in the fusion peptide. This suggests that regions of the F protein other than the fusion peptide can influence fusion activity.

These observations underline that much more information is needed on the physical interaction between the F protein and the cell membrane at the moment of fusion. Although it is probable that the fusion peptide makes the initial contact with the target membrane, it is difficult to imagine how it participates in the fusion mechanism, which necessitates the formation of a fusion pore. Viral fusion peptides have been suggested to be "sided" insertional helices, in which bulky hydrophobic amino acids are found on one face of the helix (WHITE 1990). However, GALLAHER et al. (1992) have pointed out that, in several cases, hydrophobic amino acids are not exclusively on one face of the helix and helix-breaking amino acids are also often present. They suggest that, rather than existing as a sided structure, the entire fusion peptide sequence is in contact with lipid. Thus, the fusion peptide would appear to be ideal for the initial disruption of the target membrane, but less than ideal for participation in the process of aggregation, which can be assumed to be the basis of fusion pore formation.

Using mutagenesis studies to search for regions in the F protein that contribute to its oligomeric state, we found that the leucine zipper motif (Fig. 2) is not required for tetramerization but is essential for fusion (BUCKLAND et al. 1992). In our model for MV fusion, the leucine zipper serves to pack together the α-helices in which it appears, thus forming a water-tight fusion pore which expands as more tetrameric F proteins are recruited. Leucine not only has a propensity to occur in α-helices, but, as it possesses a long branched hydrophobic side chain, it is ideal for packing them together (CRICK 1953). Isoleucine has a very similar side chain and can replace leucine in leucine zipper motifs. Support for this model has come from studies on the leucine zipper present in the envelope protein of HIV. DUBAY et al. (1992) mutagenized the isoleucine which occurs at the center of the zipper found in gp41: none of the substitutions of this single amino acid affected the oligomeric state of gp41, but certain changes abrogated fusion. When the isoleucine was replaced by leucine or valine, fusion was not affected, but changes to glycine or serine or charged amino acids inhibited fusion. Interestingly, mutagenesis to alanine gave an intermediate phenotype in which fusion was reduced. In the context of the model we have proposed, these data are very supportive. The packing of the α-helices should not be greatly affected by the conservative substitutions of leucine, isoleucine and valine, as all possess long branched hydrophobic side chains. Alanine, however, although hydrophobic, has a very short side chain; thus, it is unlikely to pack so efficiently and this could give rise to a "leaky" pore. If the model is correct, the substitution of isoleucine by charged amino acids or serine or particularly glycine would severely disrupt this packing. If the fusion proteins of these two viruses function via a common mechanism, any information derived from MV fusion could well be useful in the development of antivirals against HIV.

5 The Matrix Protein

The MV genomic and antigenomic RNAs are always encapsidated by the nucleoprotein (N) to form the helical nucleocapsids (NCs); more loosely associated are the phosphoprotein (P) and the large polymerase (L), yielding altogether the active NCs which are also packaged into virions (see S.A. HORIKAMI and S.A. MOYER, this volume). The NCs accumulate in the cytoplasm, whereas the viral envelope proteins migrate to the plasma membrane. The function of the matrix (M) protein is to bring about the unification of these two viral entities. This event could occur in one of two possible ways: (1) attachment of the M protein to the NC could lead to a change in its conformation, resulting in exposure of regions which could then interact with the cytoplasmic tails of the envelope glycoproteins, or (2) conversely, the interaction of the M protein with the envelope glycoproteins at the membrane could expose sites on the M protein, making an interaction with the NC possible. As well as being very hydrophobic, which probably reflects its interaction with the plasma membrane, the M protein is highly basic, suggesting that interaction with the NC could be via salt bridges with the acidic nucleoprotein.

Chemical cross-linking studies on purified Sendai virus have shown that the M protein can be bound to the N but not to glycoproteins (MARKWELL and FOX 1980). In vitro, the MV M protein binds to the NC (HIRANO et al. 1993). Comparison of the M protein from functional and defective SSPE MV strains showed that the latter does not bind to NCs. Analysis of chimeric M proteins indicates that mutations in the NH_2-terminal, COOH-proximal or COOH-terminal region of the M protein all abrogate NC binding, suggesting that the conformation of M is important for interaction with the NC (HIRANO et al. 1993). In a study of three SSPE MV strains the M proteins did not react with antibodies against the native M but only against the denatured antigen (WONG et al. 1991). Given the strict structural requirement for M function, conformational changes in the M protein could contribute to a nonproductive infection.

The M protein, either in MV-infected or cDNA-transfected cells, is distributed throughout the cell (GIRAUDON et al. 1984; HUBER et al. 1991; BEAUVERGER et al. 1993). Although no direct evidence has been obtained for interaction of the M protein with the viral glycoproteins, SANDERSON et al. (1993) have shown that the M protein of Sendai virus is associated with the viral glycoproteins prior to their expression at the cell surface. Furthermore, 40% of the M protein cofractionated with membranes containing the viral glycoprotein. Thus, they suggest that the M protein becomes associated with the viral glycoprotein during transport to the membrane. Further studies are needed to confirm this hypothesis.

A comparison of the M proteins of MV and CDV shows that the COOH-terminal thirds have extensive homology (BELLINI et al. 1986). Within this region are conserved segments which from their nonpolar hydrophobic and β-sheet nature probably interact with the lipid of the cell membrane. As the M protein cannot be chemically linked to the glycoproteins (MARKWELL and FOX 1980), it is unknown if it attaches to one or both viral envelope proteins. Comparison of the amino acid sequences of morbillivirus F cytoplasmic tails shows that the last 14 amino acids are highly conserved suggesting an essential function (BARRETT et al. 1987; BUCKLAND et al. 1987; TSUKIYAMA et al. 1988). Interestingly, analysis of F proteins from a number of defective SSPE MV strains showed extensive mutations in this region, half of which resulted in deletions. Such deleted F proteins are quite efficiently transported and undergo fusion (SCHMID et al. 1992; CATTANEO and ROSE 1993). Thus, truncation of part of the F protein intracellular domain may reduce functional interactions with the M protein and/or NC and thus impede efficient viral assembly and budding.

As a working hypothesis for the interaction of the M protein with the MV glycoproteins, it could thus be proposed that M interacts with the cytoplasmic tail of the F protein. As the F and H proteins appear to have a specific interaction, it would not be necessary that M and H interact directly, although this cannot be ruled out. In favor of this hypothesis, the predicted amino acid sequences of the cytoplasmic regions of the morbillivirus H protein, in contrast to those of the F protein, are not conserved (KÖVAMEES et al. 1991a).

6 Conclusions

In this review, seeking to identify the different functional activities of the envelope-associated MV proteins we have emphasised the importance of the interactions of these individual proteins, either with themselves or with other proteins. The recent identification of the MV receptor, together with the advances being made on the molecular basis of viral fusion, should provide the basis of further interest in defining the biological functions of envelope-associated MV proteins.

Acknowledgment. We would like to thank BERNADETTE MARET for invaluable editorial assistance.

References

Alkhatib G, Richardson C, Shen S-H (1990) Intracellular processing, glycosylation, and cell-surface expression of the measles virus fusion protein (F) encoded by a recombinant adenovirus. Virology 175: 262–270
Barrett T, Clarke DK, Evans SA, Rima BK (1987) The nucleotide sequence of the gene encoding the F protein of canine distemper virus: a comparison of the deduced amino acid sequence with the other paramyxoviruses. Virus Res 8: 373–376
Beauverger P, Buckland R, Wild TF (1993) (to be published) Establishment and characterisation of murine cells constitutively expressing the fusion, nucleoprotein and matrix proteins of measles virus. J Virol Methods
Bellini WJ, Englund G, Richardson CD, Rozenblatt S, Lazzarini RA (1986) Matrix genes of measles virus and canine distemper virus: cloning, nucleotide sequences, and deduced amino acid sequences. J Virol 58: 408–416
Buckland R, Wild TF (1989) Leucine zipper motif extends. Nature 547: 6216
Buckland R, Gerald C, Barker R, Wild TF (1987) Fusion glycoprotein of measles virus: nucleotide sequence of the gene and comparison with other paramyxoviruses. J Gen Virol 68: 1695–1703
Buckland R, Malvoisin E, Beauverger P, Wild TF (1992) A leucine zipper structure present in the measles virus fusion protein is not required for its tetramerization but is essential for fusion. J Gen Virol 73: 1703–1707
Cattaneo R, Rose JK (1993) Cell fusion by the envolope glycoproteins of persistent measles viruses wich caused lethal human brain disease. J Virol 67: 1493–1502
Citovsky V, Yanai P, Loyter A (1986) The use of circular dichroism to study conformational changes induced in Sendai virus envelope glycoproteins. A correlation with the viral fusogenic activity. J Biol Chem 261: 2235–2239
Crick FHC (1953) The packing of α-helices: simple coiled coils. Acta crystallographica 6: 689–697
Crise B, Buonocore L, Rose JK (1990) CD4 is retained in the endoplasmic reticulum by the human immunodeficiency virus type 1 glycoprotein precursor. J Virol 64: 5585–5593
Dubay J, Roberts S, Brody B, Hunter E (1992) Mutations in the leucine zipper of the HIV-1 transmembrane glycoprotein affect fusion and infectivity. J Virol 66:4748–4756
Ewbank J, Creighton T (1991) The molten globule protein conformation probed by disulphide bonds. Nature 350: 518–520
Fujinami Rs, Oldstone MBA (1981) Failure to cleave measles virus fusion protein in lymphoid cells. A possible mechanism for viral persistence in lymphocytes. J Exp Med 154: 1489–1499
Gallaher W, Segrest J, Hunter E (1992) Are fusion peptides really "sided" insertional helices? Cell 70: 531–532
Garcia JV, Miller D (1991) Serine phosphorylation-independent downregulation of cell-surface CD4 by nef. Nature 350: 508–511
Gerald C, Buckland R, Barker R, Freeman G, Wild TF (1986) Measles virus haemaglutinin gene: cloning, complete nucleotide sequence analysis and expression in COS cells. J Gen Virol 67: 2695–2703
Gething MJ, White JM, Waterfield MD (1978) Purification of the fusion protein of Sendai virus: analysis of the NH_2-terminal sequences generated during precursor activation. Proc Natl Acad Sci USA 75: 2737–2740

Giraudon P, Wild TF (1985) Correlation between epitopes on hemagglutinin of measles virus and biological activities: passive protection by monoclonal antibodies is related to their hemagglutination inhibiting activity. Virology 144: 46–58
Giraudon P, Gerald C, Wild TF (1984) A study of measles virus antigens in acutely and persistently infected cells using monoclonal antibodies: differences in the accumulation of certain viral proteins. Intervirology 21: 110–120
Giraudon P, Jacquier MF, Wild TF (1988) Antigenic analysis of African measles virus fields: identification and localisation of one conserved and two variable epitope sites on the NP protein. Virus Res 18:137–152
Gotoh B, Ohnishi Y, Inocencio NM, Esaki E, Nakayama K, Barr PJ, Thomas G, Nagai Y (1992) Mammalian subtilisin-related proteinases in cleavage activation of the paramyxovirus fusion glycoproteins: superiority of Furin/PACE to PC2 or PC1/PC3. J Virol 66:6391–6397
Graves MC, Silver SM, Choppin PW (1978) Measles virus polypeptide synthesis in infected cells. Viriology 86: 254–263
Hirano A (1992) Subacute sclerosing panencephalitis virus dominantly interferes with replication of wild-type measles in a mixed infection implication for viral persistence. J Virol 66: 1891–1898
Hirano A, Ayata M, Wang AH, Wong TC (1993) Functional analysis of matrix proteins expressed from cloned genes of measles virus variants that cause subacute sclerosing panencephalitis reveals a common defect in nucleocapsid binding. J Virol 67: 1848–1853
Hoxie JA, Alpers JD, Rakowski JL, Huebner K, Haggarty BS, Cedarbaum AJ, Reed JC (1986) Alterations in T4 (CD4) protein and mRNA synthesis in cells infected with HIV. Science 234: 1123–1127
Hsu MC, Scheid A, Choppin PW (1981) Activation of the Sendai virus fusion protein (F) involves a conformational change with exposure of a new hydrophobic region. J Biol Chem 256: 3557–3563
Hu X, Ray R, Compans RW (1992) Functional interactions between the fusion protein and hemagglutinin-neuraminidase of human parainfluenza viruses. J Virol 66: 1528–1534
Hu A, Sheshberadaran H, Norrby E, Kövamees J (1993) Molecular characterization of epitopes on the measles virus hemagglutinin protein. Virology 192: 351–354
Huber M, Cattaneo R, Spielhofer P, Örvell C, Norrby E, Messerli M, Perriard JC, Billeter MA (1991) Measles virus phosphoprotein retains the nucleocapsid protein in the cytoplasm. Virology 185: 299–308
Hull JD, Krah DL, Choppin PW (1987) Resistance of a measles virus mutant to fusion inhibitory oligopeptides is not associated with mutations in the fusion peptide. Virology 159: 368–372
Kövamees J, Blixenkrone-Möller M, Sharma B, Örvell C, Norrby E (1991a) The nucleotide sequence and deduced amino acid composition of the haemagglutinin and fusion proteins of the morbillivirus phocid distemper virus. J Gen Virol 72: 2959–2966
Kövamees J, Blixenkrone-Möller M, Sharma B, Örvell C, Norrby E (1991b) The nucleotide and predicted amino acid sequence of the attachment protein of canine distemper virus. Virus Res 19: 223–234
Krah DL, Choppin PW (1988) Mice immunized with measles virus develop antibodies to a cell-surface receptor for binding virus. J Virol 62: 1565–1572
Liszewski MK, Post TW, Atkinson JP (1991) Membrane cofactor protein (MCP or CD46): newest members of the regulators of complement activation gene cluster. Annu Rev Immunol 9: 431–455
Mäkelä MJ, Salmi AA, Norrby E, Wild TF (1989) Monoclonal antibodies against measles virus hemagglutinin react with synthetic peptides. Scand J Immunol 30: 225–231
Markwell MAK, Fox CF (1980) Protein-protein interaction within paramyxoviruses identified by native disulfide bonding or reversible chemical cross-linking. J Virol 33: 152–166
Morrison T, Portner A (1991) Structure, function and intracellular processing of the glycoproteins of paramyxoviridae. In: Kingsberg DW (ed) The Paramyxoviruses, Plenum, pp 347–382
Moscona A, Peluso RW (1991) Fusion properties of cells persistently infected with human parainfluenza virus type 3: participation of hemagglutinin-neuraminidase in membrane fusion. J Virol 65: 2773–2777
Mottet G, Portner A, Roux L (1986) Drastic immunoreactivity changes between the immature and mature forms of Sendai virus HN and F0 glycoproteins. J Virol 59: 132–141
Naniche D, Wild TF, Rabourdin-Combe C, Gerlier D (1992) A monoclonal antibody recognises a human cell surface glycoprotein involved in measles binding. J Gen Virol 73:2617–2624
Naniche D, Wild TF, Rabourdin-Combe C, Gerlier D (1993a) Measles virus haemagglutinin induced down regulation of gp57/67, a molecule involved in virus binding. J Gen Virol 74: 1073–1079
Naniche D, Varior-Krishnan G, Cervoni F, Wild TF, Rossi B, Rabourdin-Combe C, Gerlier D (1993b) Human membrane cofactor protein (CD46) acts as a cellular receptor for measles virus. J Virol
Neyt C, Geliebter J, Slaoui M, Morales D, Meulemans G, Burny A (1989) Mutations located in both F1 and F2 subunits of the Newcasle disease virus fusion protein confer resistance to neutralization with monoclonal antibodies. J Virol 63: 952–954

Norrby E, Sheshberadaran H, Mc Cullough KC, Carpernter WC, Örvell C (1985) Is rinderpest virus the archevirus of the morbillivirus genus? Intervirology 23: 228–232
Novick N, Hoekstra D (1988) Membrane penetration of Sendai virus glcoprotein during the early stages of fusion with liposomes as determined by hydrophobic photoaffinity labelling. Proc Natl Acad Sci USA 85: 7433–7437
Ogura H, Sato H, Kamiya S, Nakamura S (1991). Glycosylation of measles virus haemagglutinin protein in infected cells. J Gen Virol 72: 2679–2684
Portner A, Scroggs R, Naeve C (1987) The fusion glycoprotein of Sendai virus: sequence analysis of an epitope involved in fusion and virus neutralization. Virology 157: 556–559
Richardson CD, Scheid A, Choppin PW (1980) Specific inhibition of paramyxovirus and myxovirus replication by oligopeptides with amino acid sequences similar to those at the N-termini of the F1 or HA_2 viral polypeptides. Virology 105: 205–222
Rota JS, Hummel KB, Rota PA, Bellini WJ (1992) Genetic variability of the glycoprotein genes of current wild-type measeles isolates. Virology 188: 135–142
Sanderson CM, McQueen NL, Nayak DP (1993) Sendai virus assembly: M protein binds to viral glycoproteins in transit through the secretory pathway. J Virol 67: 651–663
Sato TA, Kohama T, Sugiura A (1988) Intracellular processing of measles virus fusion protein. Arch Virol 98: 39–50
Schmid A, Spielhoter P, Cattaneo R, Baczko K, ter Meulen V, Billeter MA (1992) Subacute sclerosing panencephalitis is typically characterized by alterations in the fusion protein cytoplasmic domain of the persisting measles virus. Virology 188: 910–915
Sechoy O, Philippot JR, Bienvenue A (1987) F protein-F protein interaction within the Sendai virus identified by native bonding or chemical cross-linking. J Biol Chem 262: 11519–11523
Sheshberadaran H (1991) Exposure to acidic pH causes irreversible conformational changes in the measles virus hemagglutinin. Arch Virol 117: 305–311
Sheshberadaran H, Norrby E (1986) Characterization of epitopes on the measles virus hemagglutinin. Virology 152: 58–65
Sheshberadaran H, Norrby E, McCullough KC, Örvell C (1985) The antigenic relationship between measles, canine distemper and rinderpest viruses studied with monoclonal antibodies. J Gen Virol 67: 1381–1392
Sugiyama M, Minamoto N, Kinjo T, Hirayama N, Asano K, Tsukiyama-Kohara K, Yoshikawa Y, Yamanouchi K (1991) Antigenic and functional characterization of rinderpest virus envelope proteins using monoclonal antibodies. J Gen Virol 72: 1863–1869
Taylor J, Pincus S, Tartaglia J, Richardson C, Alkhatib G, Briedis D, Appel M, Norton E, Paoletti E (1991) Vaccinia virus recombinants expressing either the measles virus fusion or hemagglutinin glycoprotein protect dogs against canine distemper virus challenge. J Virol 65: 4263–4274
Toyoda T, Gotoh B, Sakaguchi T, Kida H, Nagai Y (1988) Identification of amino acids relevant to three antigenic determinants of the fusion protein of Newcastle disease virus that are involved in fusion inhibitation and neutralization. J Virol 62: 4427–4430
Tsukiyama K, Sugiyama M, Yoshikawa Y, Yamanouchi K (1987) Molecular cloning and sequence analysis of the rinderpest virus mRNA encoding the hemagglutinin protein. Virology 160:48–54
Tsukiyama K, Yoshikawa Y, Yamanouchi K (1988) Fusion glycoprotein (F) of the rinderpest virus: entire nucleotide sequence of the F mRNA and several features of the F protein. Virology 164: 523–530
Vialard J, Lalumière M, Vernet T, Briedis D, Alkhatib G, Henning D, Levin D, Richardson C (1990) Synthesis of the membrane fusion and hemagglutinin proteins of the measles virus, using a novel baculovirus vector containing the β-galactosidase gene. J Virol 64: 37–50
White J (1990) Viral and cellular fusion proteins. Annu Rev Physiol 52: 675–697
Wild TF, Malvoisin E, Buckland R (1991) Measles virus: both the haemagglutinin and the fusion glycoproteins are required for fusion. J Gen Virol 72:439–442
Wilson I, Cox N (1990) Structural basis of immune recognition of influenza virus hemagglutinin. Annu Rev Immunol 8: 737–771
Wong TC, Ayata M, Ueda S, Hirano A (1991) Role of biased hypermutation in evolution of subacute sclerosing panencephalitis virus from progenitor acute measles virus. J Virol 65: 2191–2199
Zerial M, Huylebroeck D, Garoff H (1987) Foreign transmembrane peptides replacing the internal signal sequence of transferrin receptor allow its replication and membrane binding. Cell 48: 147–155

Measles Virus Strain Variations

B.K. Rima[1], J.A.P. Earle[1], K. Baczko[2], P.A. Rota[3], and W.J. Bellini[3]

1 Introduction

Measles virus is monotypic in nature, i.e. only a single serotype of the virus has been described. Thus, until recently, little emphasis has been placed on variation between MV strains, except for those associated with a rare neurological complication, subacute sclerosing panencephalitis (SSPE), associated with a persistent MV infection. In this chapter, we review variations between MV strains. Biological differences as well as biochemical and nucleotide sequence differences have been documented, but so far the latter have not been correlated with any specific phenotype of the virus. The structure and gene expression strategy of the virus are reviewed elsewhere in this volume (see chapter by Horikami and Moyer, this volume).

[1] School of Biology and Biochemistry, The Queen's University of Belfast, Medical Biology Centre, 97 Lisburn Road, Belfast BT9 7BL, Northern Ireland, UK
[2] Institute of Virology, University of Würzburg, Versbacher Strasse 7, 97078 Würzburg, Germany
[3] Measles Virus Section CID, Division of Viral and Rickettsial Diseases, REVG Mailstop G17, 16000 Clifton Road, Atlanta, GA 30333, USA.

2 Biological Variation Between Strains

It has long been known that MV like many other RNA viruses, can be adapted to growth in a particular host cell type. Virus stocks grown in one cell type are often unable to grow efficiently in other cells types. The molecular biological changes underlying this process are not clear. The block in the growth of the virus appears not to be related to attachment and penetration, as a wide variety of cell types can be infected with the virus. The adaptation to specific host cells probably relates to nucleotide changes or changes in the "quasi" species of RNA molecules in the population (HOLLAND et al. 1982). Adaptation of the virus by rodent brain passage, for example, has given rise to virus stocks that can grow successfully in the CNS of mice and rats after intracerebral injection, even though other virus strains are unable to do so. The nucleotide changes associated with this adaptation have not yet been elucidated. The process of adaptation to other non-human host cell types has been used successfully for the development of live attenuated vaccine strains, but again the molecular basis of this process is unknown.

In the 1970s a large number of reports were published in which the ability to grow in various cell lines was compared between MV strains derived from cases of SSPE and the laboratory adapted Edmonston B strain or its derivatives. No clear picture emerged from these studies and in retrospect their value can be questioned. Some of the so-called SSPE viruses may be simply laboratory contaminants (see below). In relation to the rapid evolution of RNA viruses, it is also possible that defective interfering (DI) particles may have played a role in partly determining these differences (RIMA et al. 1977; CALAIN and ROUX 1988). Qualitative and quantitative differences in DI particles may affect the growth of MV in different cell types, with some cells (e.g. Vero cells) being more readily able to replicate them than others, for example, HeLa cells (Rima, unpublished observations). The ability to induce interferon may have been another factor that partially explained the differences between various virus stocks. MCKIMM and RAPP (1977) demonstrated that the ability to induce interferon is lost rapidly when virus stocks usually propagated on Vero cells and able to induce interferon in BSC1 cells, are propagated on BSC1 cells. Readaptation to Vero cells restored the ability to induce interferon quickly. Whether this was related to the presence of DI particles, which in other systems have been shown to have a role in interferon induction, has never been determined.

Some 33 temperature sensitive (ts) mutants of MV have been described (BERGHOIZ et al. 1975; HASPEL et al. 1975), but unfortunately these have not been analysed extensively as they had very high frequencies of reversion. None were characterised in molecular terms, even though at least one showed an interesting phenotype in its ability to increase the frequency of hydrocephalus in hamsters (BRESCHKIN et al. 1976). Cold sensitive mutants of MV have also been described but these have not been characterised further (RAGER-ZISMAN et al. 1984). However, stable cold-adapted vaccine strains have been established which fail to produce viral progeny or plaques at 40°C (SUZUKI et al. 1990). Another apparently very stable ts replication-negative mutant (VYDELINGUM et al. 1989) has been obtained by 5-fluoruracil mutagenesis (CHUI et al. 1986).

In all morbilliviruses, strains with differing types of cytopathic effect have been recognized. Small and large plaque forming variants have been isolated as well as strains that form no plaques at all but lead to the formation of clumps (foci) of infected cells (RAPP 1964; GOULD et al. 1976). The fusiogenic variants have been shown to be more neurovirulent for hamsters (CARRIGAN 1986). The molecular basis of these differences has not been elucidated, but it is likely that the plaque types differ in their ability to fuse infected cells from within. Fusion from without has never been demonstrated for MV and thus, unlike for other paramyxoviruses, such as Newcastle disease virus, variation in this property cannot be assessed for MV. Strains with different abilities to fuse cells from within have been described. The mechanisms involved in fusion, shown to require concerted action by the hemagglutinin and fusion proteins (WILD et al. 1991), are being studied actively at present, as several groups are attempting to determine the nature of viral receptor(s). In the F protein of all paramyxoviruses, a hydrophobic peptide sequence present at the NH_2-terminal of the F1 part of the protein has been proposed to be responsible for the fusion of the viral membrane with that of the cell. Oligopeptides with the same sequence are able to inhibit fusion by MV (RICHARDSON et al. 1980). Mutants that are resistant to these inhibitors have been isolated and sequenced but the resistance is not associated with mutations in the region encoding this peptide sequence but with mutations elsewhere in the molecule (HULL et al. 1987). It is possible that these mutations exert their effect through alterations in the conformation of the F protein.

Recent studies have identified human membrane cofactor protein (CD46) (LISEWSKI and ATKINSON 1992) as a receptor for measles virus (NANICHE et al. 1993; confirmed by DÖRIG et al. 1993). Studies are still at an early stage and whether different strains of measles vary in their ability to bind CD 46 is not known. Some morbilliviruses such as MV have the ability to agglutinate red blood cells from humans or selected species of animals like many but not all members of the paramyxovirus family. MV only hemagglutinates monkey and some human red blood cells. Variant viruses have been described that only agglutinate red blood cells in the presence of high salt concentrations (0.8 M ammonium sulphate). These so-called salt-dependent hemagglutinin variants have been found to be present in a number of different MV isolates (GOULD et al. 1976; SHIRODARIA et al. 1976). The molecular basis of the phenomenon is not known, but our recent results (S. FLANAGAN and B.K. RIMA, unpublished) suggest that this may be associated with a mutation in the H protein, 117F→L. The hemagglutinins from some recent wild-type strains have very low hemagglutination activity (SAITO et al. 1992) which might be caused by the presence of an additional N-linked glycosylation site at amino acid 416 of the H-protein probably correlating with its retarded SDS-PAGE migration (SAKATA et al. 1993).

Although MV is monotypic, differences in the presence of specific epitopes, defined by the binding of monoclonal antibodies to the virus, have been described. In 1983, SHESHBERADARAN and coworkers (1983) demonstrated that a number of strains differed in their patterns of monoclonal antibody binding. It was found in this study that the M and H protein had the widest degree of variation between strains whilst the P, N and F proteins were antigenically more stable. This study has been followed up with others analysing specific MV antigens in more detail. A study on the N protein confirmed the presence of epitopic variation in the virus (GIRAUDON et al. 1988).

3 Analytical Differences of Measles Virus Proteins

Variations in the mobility of the various viral proteins in SDS-PAGE were studied extensively in the 1970s. Differences were reported in the (apparent) sizes of the H, P and M proteins. Variations in the SDS-PAGE migration of the H protein have also been found in recently described wild-type viruses (ROTA et al. 1992; SAITO et al. 1992; SAKATA 1993). These H molecules have an additional potential N-linked glycosylation site, but it has not yet been unequivocally determined that this new site is utilized. Some of the variation in the mobility of the P proteins of various MV strains is related to an as yet unidentified form of post translational modification and to amino acid sequence differences. The molecular mass of the P protein derived form the primary sequence is approximately 51 kDa (BELLINI et al. 1985); however, the protein migrates with an apparent mass of approximately 70 kDa. Undoubtedly, phosphorylation plays a major role in the reduced migration through the gel, but the slow and fast migrating forms in doublet bands in SDS-PAGE have not been associated with differences in phosphorylation levels. There seem to be differences between strains in the different levels of these two bands. The doublet is not sequence-specific because by plaque purification it has not been possible to resolve strains bearing either the slow or fast forms of the P protein. Sequence differences, occasionally reflecting a single amino acid change, can have substantial effects on the migration of proteins in SDS-PAGE. Nucleotide sequence differences found in the genes encoding the P, C and V proteins might explain some of the observed variation in migration (CATTANEO et al. 1989).

The mobility of the matrix protein of MV has been studied extensively in relation to its potential significance as a marker for SSPE (WECHSLER and FIELDS, 1978). However, the differences found turned out not to be characteristic for SSPE (RIMA et al. 1979). A number of wild-type viruses appear to have an M protein with a slower mobility (39 kDa instead of 37 kDa) and this is a stable characteristic of these strains.

In summary, a number of strain differences have been described for MV but none of these have as yet been found to be associated with a specific phenotype or have been explained in terms of nucleotide sequence differences.

4 Genetic Variation and Evolution of the Virus

The observation that MV strains varied in their ability to bind to monoclonal antibodies indicates the existence of genetic variation in this monotypic virus. In 1989, it became clear that there were strain-specific changes in nucleotide sequence besides those that were associated with prolonged brain infections by the virus (CATTANEO et al. 1989). This study indicated for the first time that changes in the COOH-terminal 126 amino acid residues of the N gene and areas in the P and H genes did not affect the functionality of the proteins but reflected strain variation. The original descriptions of the comparison of canine distemper virus (CDV) with MV had already indicated that certain areas of the genome of morbilliviruses have greater plasticity than others (ROZENBLATT et al. 1985; BELLINI et al. 1986). Comparisons of the sequences published before 1989 indicated that the region of the genome between the open reading frames (ORFs) encoding the M and F proteins is GC rich and varies substantially in related strains (BELLINI et al. 1986; RICHARDSON et al. 1986; BUCKLAND et al. 1987). However, as this region is presumed to be non coding, it is less interesting to study, and the difficulties encountered in sequencing this region, which has strong secondary structure determinants, led us to concentrate on other parts of the genome. From comparisons and alignments of the major structural proteins of paramyxo- and morbilliviruses it has become clear (RIMA 1989) that the regions encoding the COOH-terminal part of the N protein and the NH_2-terminal 100 amino acids of the P and C proteins (BACZKO et al. 1992) are variable parts of the genomes of this group of viruses. A somewhat lower degree of variation was found in the H proteins. The H and N protein coding sequences have been analysed in great detail, primarily because these two proteins elicit strong humoral immune responses and the N protein may be involved in the generation of T cell-mediated immunity.

The sequence of the COOH-terminal 151 amino acids of the N protein has now been analysed in 46 strains of MV. It shows that there is up to 7.2% divergence in the coding nucleotide sequence and 10.6% divergence in the amino acid sequence between the most unrelated strains in this region (TAYLOR et al. 1991). MV strains fall into six or seven different genotypes some of which are extinct (i.e., have not been isolated recently) and others which are still cocirculating in the human population (TAYLOR et al. 1991). In this respect, MV does not differ from other paramyxoviruses such as human respiratory syncytial virus (CANE and PRINGLE 1991), human parainfluenza virus type 3 (VAN WYKE COELINGH et al. 1988) and mumps virus (YEO et al. 1993). The grouping of strains in genotypes is shown in Table 1 and displayed as an unrooted tree in Fig. 1.

The first group contains all of the vaccine strain used so far. The type strain is the Edmonston strain of MV and the relationships within the group are described below in greater detail. It also contains early American isolates such as the Philadelphia strain and a wild-type isolate from Turku (Finland) in 1962 (P. A. ROTA et al. 1994). The group also contains the Hallé strain, a supposedly SSPE-derived

A second fact emerging from this analysis, and the one shown in Fig. 1, is that not in a single case was an isolate found which was a direct precursor for any of the later viruses. The overall data in the diagram indicate that mutations accumulate, and later isolates in the same group are placed further to the right (increasing distances from the root) but none of the isolates is on a node. This indicates the fragmentary nature of the record that we have of MV evolution due to infrequent isolation and highlights the differences with studies on the evolution of influenza A virus.

Analysis of both the H protein and the N gene indicates that the fifth lineage group is very diverse and can be considered to be split into two subgroups. Although some of the UK isolates from the early 1970s and the Northern Ireland isolates from the early 1980s form a specific subgroup, they share a large number of mutations with the sequences from the later UK and the recent USA isolates as well as 1988 isolates from Japan (SAKATA et al. 1992). Members of the fifth very diverse genotype were thus isolated in the early 1970s in the UK, the early 1980s in the USA (Bo and BE83) and later in Japan and in the 1989-1991 outbreak in the USA. It is a diverse group, particularly since the more recent isolates have a new H protein glycosylation site (416D→N) (ROTA et al. 1992). This site is close to the variable site identified in monoclonal antibody (Mab) escape mutants by HU et al. (1993) and LIEBERT et al. (1994). Whether these changes in glycosylation alter the accessibility of the surface to antibodies or whether conformational changes occur is not known, as the Mabs used for the escape mutant generation recognize conformational rather than linear epitopes. This is not the only example of the change of glycosylation sites as the one at position 200 is changed in the rat adapted CAM strain of virus (200N→R) and a corresponding mutation occurred in one of the SSPE cases (CATTANEO et al. 1989). Generally the additional glycosylation sites are defined only by the amino acid sequence; the actual presence of oligosaccharides at the suggested sites remains open.

Whether these lineage groups are tabulated as shared mutations (TAYLOR et al. 1991) or whether their relationships are shown as unrooted trees (Figs. 1,2) it is clear that the groupings are consistent and independent of the area of the genome that is compared. The relationships between strains derived from comparisons of the H, P, M and N genes are essentially the same (TAYLOR et al. 1991; BACZKO et al. 1991, 1992; ROTA et al. 1992). This indicates the absence of recombination between these lineage groups over the time period studied. Since MV is hightly infectious and since vigorous immune responses are associated with infection, the window of opportunity for coinfection with two virus strains from different linage groups and thus the opportunity to recombine is probably very small. There is no evidence for recombination in any of the mononegavirales.

The analysis has also indicated that the sequence of the SBI isolate from Germany in 1992 is outside the other groups and may represent a new lineage group. The sequence of the N gene also supports this (RIMA et al., in preparation). It will be interesting to see whether this strain is related to lineage groups circulating elsewhere which have not yet been identified.

5 Mutations Associated with Attenuation and Comparisons of Vaccine Strains

A comparison of the nucleotide sequences of all the currently available vaccines indicates that they all belong to the same genotype as the Edmonston strain. Figure 3 shows an unrooted tree of the relationships of the vaccine strains (group 1) analysed in contrast to the recent wild-type isolate Chi (group 5) on the basis of the N, F and H genes (J.S. ROTA et al. 1994). All the strains isolated before 1971 appear to belong to this single group 1 genotype, including strains isolated over a very wide geographical area, i.e. the Edmonston and Philadelphia strains from the USA in 1954 and 1957, respectively, the Chanchung strain in 1957 in China, the Leningrad strain in 1960 in Russia, the Halonen strain from Finland in 1962 and the CAM (Tanabe) strain in 1968 in Japan. However, after 1971 only wild types with other genotypes have been isolated and, although variations occur between the strains, these are not significantly greater than those found in other groups. Vaccination with the Edmonston-derived vaccines was initiated in many countries much later than 1971. This makes it unlikely that this immunization programme has led to the rapid disappearance of the world-wide Edmonston genotype and its replacement by five or more different genotypes. Some of the new genotypes have been in circulation for at least 15 years already and the rate of accumulation of mutations indicates that these represent long established genotypes of MV. The situation for MV is also in stark contrast to that for another human childhood paramyxovirus infection, mumps, in which a number of very distinct geographically restricted genotypes have been described (YEO et al. 1993). As to the question of the biological importance of the fact that all vaccines have a genotype

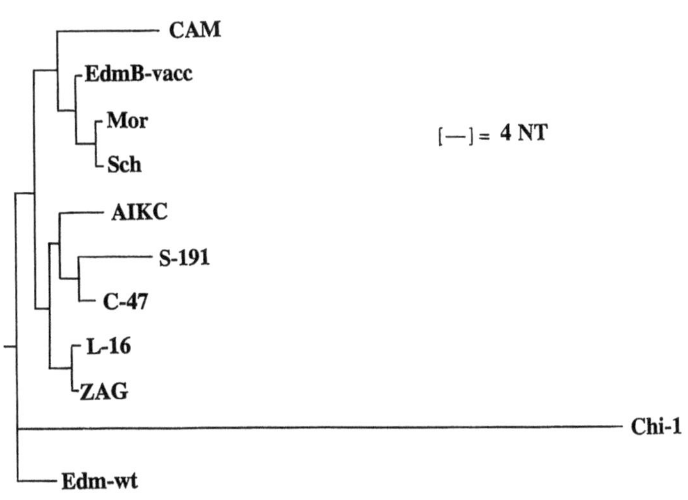

Fig. 3. Unrooted tree of the vaccine strains derived from the H, F and N gene sequences fo several viruses. Chi1 strain belongs to another genotype. *Horizontal bar* represents four nucleotide changes.

substantially different from the currently circulating wild-type viruses, there is no indication that the currently used vaccines are not able to control MV infection with viruses of differing genotypes. However, this could have an effect on the length of the window of opportunity of infection of children with waning maternal antibody responses generated either by vaccination with the Edmonston strain or by infection with wild-type viruses with a different genotype.

At present, attempts are being made to analyse the mutations that are associated with the attenuated phenotype of MV vaccines. However, due to the fact that only one direct vaccine and progenitor pair is available for the Edmonston strain and that the original, low passage number, Edmonston wild-type strain is no longer available, it is difficult to arrive at conclusions about the significance of mutations on the basis of comparative nucleotide sequence analysis of vaccine and wild-type strains. Mori et al (1993) sequenced the entire genome of the AIK-c strain of measles, an Edmonston-derived vaccine. The sequence of the Edmonston strain has been compiled (Galinski 1991). The AIK-c strain had 56 nucleotide and 31 amino acid changes compared to the Edmonston strain. We are currently sequencing the wild-type virus to get a list of mutations that differ from the vaccine strains. So far, very few mutations have been observed in the genes encoding some of the structural proteins of MV (Rota et al. 1994; Rima and Earle, in preparation) and one is struck by the remarkable stability of the nucleotide sequence of MV strains. However, it cannot be excluded that a small number of mutations in control regions, as in the case of poliovirus, may be responsible for the attenuated phenotype. It is also clear that at least in the intracerebral infection of Lewis rats Mab escape mutations of the H protein can affect the neurovirulence of the rat brain adapted CAM strain of virus (Liebert et al. 1994).

6 Mutations in Subacute Sclerosing Panencephalitis and Measles Inclusion Body Encephalitis

The study of MV strains associated with SSPE and measles inclusion body encephalitis (MIBE) so far has involved analysis of virus isolates from the CNS and lymph nodes of the patients; of persistently infected cells established by cocultivation of CNS tissue with susceptible (monkey kindney) cells or of cDNAs cloned directly form RNA extracted from autopsy material. The analyses have shown that there are mutations in transcription patterns and mutations that disrupt the coding potential of viruses and that the phenomenon of biased hypermutation occurs in many nucleotide sequences (mainly of the matrix gene) obtained directly from brain material. As indicated above, there are reasons to question the significance of the two replication competent strains that have been isolated from SSPE cases and have been found to belong to the Edmonston lineage group. All the other studies have shown a fairly consistent pattern of variation in SSPE and

MIBE viruses, with an overall emphasis on reduction of the expression of the envelope-associated antigens M, F and H.

Concerning mutations affecting transcription patterns, in a number of SSPE cell lines and brains readthrough RNA transcripts have been identified in which the P-M intergenic junction has been ignored. This leads to the formation of large transcripts from which it is impossible to obtain translation of the M ORF in vitro (CATTANEO et al. 1986; WONG and HIRANO 1987). In one case, the readthrough may be related to changes in the intergenic nucleotide sequences, since the GAA intergenic sequence of the template has been mutated to GGA (YOSHIKAWA et al. 1990) but in another persistently infected cell line and in one SSPE brain there is no variation in the sequence associated with this altered pattern of recognition of the P-M intergenic region (CATTANEO et al. 1986). It is presumed that as yet unidentified changes in the L or other viral genes are responsible for the altered transcription pattern in the latter two examples.

A second simple way of reducing the expression of, for example, the M protein is by the introduction of mutations in the M ORF. In one publication (CATTANEO et al. 1989) the significance of the changes in the M gene, as compared to those in the N, P, F and H genes, was underestimated since the SSPE- and MIBE- derived MV strains were compared with a lytic consensus sequence that only contained the first lineage group. In later studies comparing SSPE-derived by Wild and Buckland, (this volume) viruses with the wild-type strains of the same lineage group (BACZKO et al. 1991) the significance of changes specifically in the M gene became more pronounced. In some cases (CATTANEO et al. 1986) introduction of an early stop codon has obliterated the potential of the virus to express the M protein which is imporant in the budding process (see chapter IX). In others, mutations in the coding sequences have been identified that lead to a greatly enhanced turnover of the protein (SHEPPARD et al. 1986; CATTANEO et al. 1988b). In these cases, revertants with increased stability of the M protein have been identified. Thus, either by stop codons or by mutations that induce instability, the effective concentration of the M protein can be obliterated or reduced in cases of SSPE.

A second class of mutation, affecting the F protein, has recently been shown to be associated with all SSPE cases analysed so far and entails the small COOH-terminal cytoplasmic region, whereas the bulk of the F protein is well conserved (SCHMID et al. 1992). Premature stop codons are the most frequent mutations, but elongation of the reading frame, frame shifts and nonconservative amino acid replacements occur. These mutations appear particularly significant since they affect the COOH-terminal 15 amino acids that are perfectly conserved in all morbilliviruses. In some cases, mutant and wild-type virus sequences have been found to co-exist. Whether both types of viruses are present in the same cells is not known.

Another form of mutation that has been observed predominantly in the M gene of many of the SSPE and MIBE cases analysed is a specific form of biased hypermutation associated probably with the presence of double-stranded RNA "unwindase" enzyme in the cell (BASS et al. 1989). The activity is present in large

established and the site of persistence of the virus and the explanation of the male predominance in affected children remain to be elucidated (Rima 1994). The pathogenesis of this disease will thus remain of great interest as a prime example of the interaction of the human host with a persisting virus, which due to its RNA genome requires constant replication, albeit probably at very low levels.

The monotypic nature of MV in serological terms has masked the existence of a set of defined genotypes, which accumulate mutations continuously. Two matters are of potential significance. First of all, it is immediately striking that there are very few silent changes found in comparisons of similar strains and the number of expressed mutations between genotypes is larger than expected. This indicates that selective forces are operating on the virus, which are not expected to favour silent mutations. Second, the main question raised by the recent data concerns the biological significance of the genotypes. To what extent this variation is driven by the immune system remains to be determined. Antigenic differences are detectable between vaccine and wild-type viruses using monoclonal and postinfection human serum (unpublished). These differences need further study as it is not clear how much this affects the ability of new wild-type viruses to replicate and be transmitted from persons with waning titres of immunity generated by vaccination with a different genotype. The significance of cell-mediated immune responses in the clearance of MV points to a potential importance of the variation in the N and to a lesser extent the H proteins of MV. This variation might be a crucial pathogenic determinant of different viral genotypes in specific individuals or in populations in which specific HLA types predominate. It is difficult to assess this at present, but it is clear that the virus is still capable of inflicting significant morbidity and mortality in developing countries. Though to a large part this may be attributed to malnourishment and hygienic conditions favouring secondary infections, there may also be virological factors that determine the significant differences in MV infection between developing and developed countries.

Acknowledgments. We would like to acknowledge support from the World Health Organisation and the Wellcome Trust for part of these studies.

References

Baczko K, Brinckmann U, Pardowitz I, Rima BK, ter Meulen V (1991) Nucleotide sequence of the genes encoding the matrix protein of two wild-type measles virus strains. J Gen Virol 72: 2279–2282

Baczko K, Pardowitz I, Rima BK, ter Meulen V (1992) Constant and variable regions of measles virus proteins encoded by the nucleocapsid and phosphoprotein genes derived from lytic and persistent viruses. Virology 190: 469–474

Baczko K, Lampe J, Liebert UG, ter Meulen V, Pardowitz I, Budka H, Cosby SL, Isserte S, Rima BK (1993) Clonal expansion of hypermutated measles virus in a SSPE brain. Virology 197: 188–195

Bass BL, Weintraub H, Cattaneo R Billeter MA (1989) Biased hypermutation of viral RNA genomes could be due to unwinding/modification of double stranded RNA. Cell 58: 33

Bellini WJ, Englund G, Rozenblatt S, Arnheiter H, Richardson CD (1985) Measles virus P gene codes for two proteins. J Virol 53: 908–919

Bellini WJ, Englund G, Richardson CD, Rozenblatt S, Lazzarini RA (1986) Matrix genes of measles virus and canine distemper virus: cloning, nucleotide sequences, and deduced amino acid sequences. J Virol 58: 408–416

Bergholz CM, Kiley MP, Payne FE (1975) Isolation and characterization of temperature-sensitive mutants of measles virus. J Virol 16: 192–202
Breschkin AM, Haspel MV, Rapp F (1976) Neurovirulence and induction of hydrocephalus with parental, mutant and revertant strains of measles virus. J Virol 18: 809–811
Buckland R, Gerald C, Barker R, Wild TF (1987) Fusion glycoprotein of measles virus: nucleotide sequence of the gene and comparison with other paramyxoviruses. J Gen Virol 68: 1695–1763
Calain P, Roux L (1988) Generation of measles virus defective interfering particles and their presence in a preparation of attenuated live virus vaccine. J Virol 62: 2859–2866
Cane PA, Pringle CR (1991) Respiratory syncytial virus heterogenuity during an epidemic: analysis by limited nucleotide sequencing (SH gene) and restriction mapping (N gene). J Gen Virol 72: 349–357
Carabana J, Caballero M, Ortego J, Fernandez-Munoz R, Celma ML (1993) Mutation patterns in measles virus envelop proteins from two subacute sclerosing panencephalitis brains of contemporary patients with the same geographical origin. Abstracts of the IXth international congress of virology, Glasgow Scotland. p 295
Carrigan D (1986) Round cell variant of measles virus: neurovirulence and pathogenesis of acute encephalitis in new born hamsters. Virology 148: 349–359
Cattaneo R, Rebmann G, Schmid A, Baczko K, ter Meulen V, Bellini WJ, Rozenblatt S, Billeter MA (1986) Accumulated measles virus mutations in a case of subacute sclerosing panencephalitis: interrupted matrix protein reading frame and transcription attenuation. Virology 154: 97–107
Cattaneo R, Schmid A, Eschle D, Baczko K, ter Meulen V, Billeter MA (1988a) Biased hypermutation and other genetic changes in defective measles viruses in human brain infections. Cell 55: 255–265
Cattaneo R, Schmid A, Billeter MA, Sheppard RD, Udem SA (1988b) Multiple viral mutations rather than host factors cause defective measles virus gene expression in a subacute sclerosing panencephalitis cell line. J Virol 62: 1388–1397
Cattaneo R, Schmid A, Spielhofer P, Kaelin K, Baczko K, ter Meulen V, Pardowitz I, Flanagan S, Rima BK, Udem SA, Billeter MA (1989) Mutated and hypermutated genes of persistent measles viruses which caused lethal human brain diseases. Virology 173: 415–425
Chui L W-L, Vainionpää R, Marusyk R, Salmi A Norrby E (1986) Nuclear accumulation of measles virus nucleoprotein associated with a temperature-sensitive mutant. J Gen Virol 67: 2153–2161
Döring RE, Marcil A, Chopra A, Richardson C (1993) The human CD46 molecule is a receptor for measles virus (Edmonston strain). Cell 75: 295–305
Ecker A, ter Meulen V, Schneider-Schaulies S (1994) Differentiation dependent regulation of unwinding/modifying activity for measles virus in neural cells (submitted for publication)
Galinski MS (1991) Annotated nucleotide and protein sequences for selected Paramyxoviridae. In: Kingsbury DW (eds) The paramyxoviruses. Plenum, New York, pp 537–588
Giraudon P, Jacquier MF, Wild TF (1988) Antigenic analysis of African measles field isolates: identification and localisation of one conserved and two variable epitope sites on the NP protein. Virus Res 10: 137–153
Gould EA, Cosby SL, Shirodaria PV (1976) Salt dependent haemagglutinating measles virus in SSPE. J Gen Virol 33: 139–142
Haspel MV, Duff R, Rapp F (1975) Isolation and preliminary characterization of temperature-sensitive mutants of measles virus. J Virol 16: 1000–1009
Holland JK, Spindler F, Horodyski E, Graham S, Nichol S, VanDePol S (1982) Rapid evolution of RNA genomes. Science 215: 1577–1585
Horta-Barbosa L, Hamilton R, Wittig B, Fucillo DA, Sever JI, Vernon ML (1971) Subacute Sclerosing Panencephalitis: isolation of suppressed measles virus from lymph node biopsy. Science 173: 840–841
Hu A, Sheshberadaran H, Norrby E, Kövamees J (1993) Molecular characterisation of epitopes on the measles virus hemagglutinin protein. Virology 192: 351–354
Hull JD, Krah DL, Choppin PW (1987) Resistance of measles virus mutant to fusion inhibitory oligopeptides is not associated with mutations in the fusion peptide. Virology 159: 368–372
Lieber UG, Flanagan SG, Löffler S, Baczko K, ter Meulen V, Rima BK (1994) Definition of antigenic determinants in the haemagglutinin of measles virus important for neurotropism and neurovirulence. J Virol 68: 1486–1493
Lisewski MK, Atkinson JP (1992) Membrane cofactor protein. In: Parker CJ (ed) Membrane defenses against attack by complement and perforins. Springer, Berlin Heidelberg New York pp 45–60 (Current topics in microbiology and immunology, vol 178)
McKimm J, Rapp F (1977) Variation in ability of measles virus plaque progeny to induce interferon. Proc Natl Acad Sci USA 74: 3056–3059
Mori T, Sasaki K, Hashimoto H, Makino S (1993) Molecular cloning and complete nucleotide sequence of genomic RNA of the AIK-c strain of attenuated measles virus. Virus Genes 7: 67–81

Naniche D, Varior-Krishnan G, Cervoni F, Wild F, Rossi B, Rabourdin-Combe C, Gerlier D (1993) Human membrane cofactor protein (CD46) acts as the cellular receptor for measles virus. J Virol 67: 6025–6032
Ragar-Zisman B, Egan JE, Kress Y, Bloom BR (1984) Isolation of cold-sensitive mutants of measles virus from persistently infected murine neuroblastoma cells. J Virol 51: 845–855
Rapp F (1964) Plaque differentiation and replication of virulent and attenuated strains of measles virus. J Bacteriol 88: 1448–1458
Rataul SM, Hirano A, Wong TC (1992) Irreversible modification of measles virus RNA in vitro by nuclear-unwinding activity in human neuroblastoma cells. J Virol 66: 1769–1773
Richardson CD, Scheid A, Choppin PW (1980) Specific inhibition of paramyxovirus and myxovirus replication by oligopeptides with amino acid sequences similar to those at the N-termini of the F1 on HA2 viral polypeptides. Virology 105: 205–222
Richardson CD, Hull D, Greer P, Hasel K, Berkovich A, Englund G, Bellini WJ, Rima BK, Lazzarini RA (1986) The nucleotide sequence of the mRNA encoding the fusion protein of measles virus (Edmonston strain): a comparison of fusion proteins from several different paramyxoviruses. Virology 155: 508–523
Rima BK (1989) Comparison of amino acid sequences of the major structural proteins of the paramyxo- and morbilliviruses. In: Kolakofsky D, Mahy BWJ (eds) Genetics and Pathogenicity of Negative Strand Viruses. Elsevier, Amsterdam, pp 254–262
Rima BK (1994) The Pathogenesis of Subacute Sclerosing Panencephalitis. Rev Med Virol 4: 81–90
Rima BK, Davidson WB, Martin SJ (1977) The role of defective interfering particles in persistent infection of Vero cells by measles virus. J Gen Virol 35: 89–97
Rima BK, Martin SJ, Gould EA (1979) A comparison of polypeptides in measles and SSPE virus strains. J Gen Virol 42: 603–608
Rota JS, Humel KB, Rota PA, Bellini WJ (1992) Genetic variability of the glycoprotein genes of current wild type measles isolates. Virology 188: 135–142
Rota JS, De-Wang Z, Rota PA, Bellini WJ (1994) Comparison of sequences of the H, F and N coding genes of measles virus vaccine strains. Virus Res 31: 317–330
Rota PA, Bloom AE, Vanchier JA, Bellini WJ (1994) Evolution of the nucleoprotein and matrix genes of wild-type strains of measles virus isolated from recent epidemics. Virology 198: 724–730
Rozenblatt S, Eizenberg O, Ben-Levy R, Lavie V, Bellini WJ (1985) Sequence homology within the morbilliviruses. J Virol 53: 684–690
Saito H, Sato H, Abe M, Harata S, Amano K, Suto T, Morita M (1992) Isolation and characterization of measles virus strains with low hemagglutination activity. Intervirology 33: 57–60
Sakata H, Kobune F, Sato TA, Tanabayashi K, Yamada A, Sigiura A (1993) Variation in field isolates of measles virus during an 8-year period in Japan. Microbiol Immunol 37: 233–237
Schmid A, Spielhofer P, Cattaneo R, Baczko K, ter Meulen V, Billeter MA (1992) Subacute sclerosing panencephalitis is typically characterized by alterations in the fusion protein cytoplasmic domain of the persisting measles virus. Virology 188: 910–915
Sheppard RD, Raine CS, Bornstein MB, Udem SA (1986) Rapid degradation restricts measles virus matrix protein expression in a subacute sclerosing panencephalitis cell line. Proc Natl Acad Sci USA 83: 7913–7917
Sheshberadaran H, Chen SN, Norrby E (1983) Monoclonal antibodies against five structural components of measles virus: I: characterisation of antigenic determinants on nine strains of measles virus. Virology 128: 341–353
Shirodaria PV, Dermott SE, Gould EA (1976) some characteristics of salt-dependent haemagglutinating measles virus. J Gen Virol 33: 107–115
Suzuki K, Morita H, Katoh M, Kidokoro M, Saika S, Yoshizawa S, Hashizume S, Horiuchi K, Okabe N, Shinozaki T, Hirayama M (1990) Development and evaluation of the TD97 measles virus vaccine. J Med Virol 32: 194–201
Taylor MJ, Godfrey E, Baczko K, ter Meulen V, Wild TF, Rima BK (1991) Identification of several different lineages of measles virus. J Gen Virol 72: 443–447
Van Wyke Coelingh KL, Winter CC, Murphy BR (1988) Nucleotide and deduced amino acid sequence of hemagglutinin-neuraminidase genes of human type 3 parainfluenza virus isolated from 1957 to 1983. Virology 162: 137–143
Vydelingum S, Ilonen J, Salonen R, Marusyk R, Salmi A (1989) Infection of human peripheral blood mononuclear cells with a temperature-sensitive mutant of measles virus. J Virol 63: 689–695
Wechsler SL, Fields BN (1978) Differences between the intracellular polypeptides of measles and subacute sclerosing panencephalitis virus. Nature 272: 458–460

Wild TF, Malvoisin E, Buckland R (1991) Measles virus: both the hemagglutinin and fusion glycoproteins are required for fusion. J Gen Virol 72: 439–442
Wong TC, Hirano A (1987) Structure and function of bicistronic RNA encoding the phosphoprotein and matrix protein of measles virus. J Virol 61: 584–589
Yeo RP, Afzal MA, Forsey T, Rima BK (1993) Identification of a new mumps virus lineage by nucleotide sequence analysis of the SH gene of ten different strains. Arch Virol 128: 371–377
Yoshikawa Y, Tsuruoka H, Matsumoto M, Haga T, Shioda T, Shibuta H, Sato TA, Yamanouchi K (1990) Molecular analysis of structural protein genes of the Yamagata-1 strain of defective subacute sclerosing panencephalitis: II nucleotide sequence of cDNA corresponding to the P plus M dicistronic mRNA. Virus Genes 4: 151–162

were depressed for 3–4 weeks following vaccination with a live attenuated but not a killed measles virus vaccine (FIREMAN et al. 1969).

The immune suppression caused by measles virus is of considerable clinical importance, as measles kills more than 1 million children worldwide per year, and most of the deaths are due to secondary infections which occur as a consequence of immune suppression (MILLER 1964; BECKFORD et al. 1985). Because of its clinical importance, much research effort has been directed at elucidating how measles virus causes immune suppression; but in spite of this, the mechanisms involved are still not fully understood. Some examples of the different types of mechanism by which viruses can produce a generalized suppression of the host immune response are shown in Table 1. Viruses may exert their suppressive effects either through an action on the effector cells of the immune response (T and B lymphocytes), and/or by affecting the functioning of accessory cells (monocyte/macrophages, dendritic cells etc.) essential for development of effector cell responses. Frequently more than one of the mechanisms listed in Table 1 may contribute to achieving the immune suppression observed following a particular virus infection in vivo as illustrated by human immunodeficiency virus type (HIV-1) infection of man, where a plethora of different mechanisms have been proposed to contribute to the development of acquired immunodeficiency syndrome (AIDS) (reviewed by WEISS 1993). Measles virus-induced immune suppression may also be multifactorial. In this chapter we will review what is known of measles virus interaction with immune system cells in vitro and in vivo, describe in vitro studies which have investigated mechanisms by which the virus may suppress immune responses and discuss how these in vitro findings may relate to the mechanics of the immune suppression seen following acute measles virus infection in vivo.

2 Measles Virus Infects Cells of the Immune System

All viruses studied to date which cause immune suppression in their hosts, indeed all viruses known to be capable of establishing persistent infections in vivo, infect cells of the immune system (reviewed by MCCHESNEY and OLDSTONE 1987). Measles virus is no exception: as long ago as 1937, Papp demonstrated that peripheral blood mononuclear cells (PBMCS) are infected during acute measles by transferring disease to naive individuals. During acute measles infection, the virus undergoes primary replication in the respiratory tract, then disseminates throughout the body and undergoes a round of secondary replication in lymphoid tissues (YAMANOUCHI et al. 1973; White and Boyd 1973). Close to the onset of the rash, there is a viremia during which infectious virus can be recovered from PBMCS but not plasma (GRESSER and CHANY 1963; PEEBLES 1967). Infectious virus/viral antigens can be demonstrated for 1 week or even more after the rash in mitogen-stimulated PBMCS (OSUNKOYA et al. 1974; WHITTLE et al. 1978;

Measles Virus-Mononuclear Cell Interactions

P. Borrow and M.B.A. Oldstone

1 Introduction

The immune system has evolved to defend the body from pathogens; and concurrently pathogens, including viruses, have developed a variety of ingenious strategies to interfere with the immune defense mechanisms and hence promote their persistence at the individual or population level. The outcome of an infection is thus dependent on a two-way interaction between the pathogen and cells of the immune system.

Measles was the infection with which the phenomenon of virus-induced immunosuppression was first demonstrated: von Pirquet (1908) observed that the tuberculin skin test response of immune individuals was transiently depressed during the course of acute measles virus infection, from just prior to the rash until 7–20 or more days following its appearance. This finding was exploited therapeutically for the treatment of nephrotic syndrome (Blumberg and Cassady 1947). Smithwick and Berkovich (1969) extended von Pirquet's initial observations in an in vitro system, demonstrating that the proliferative response to tuberculin of immune lymphocytes isolated during the course of acute measles infection is depressed. It was further shown that this suppression was actually due to the virus, as culture of lymphocytes with measles virus in vitro suppressed their proliferative response to tuberculin antigen or a mitogen (Smithwick and Berkovich 1969; Zweiman 1971). Likewise skin test responses and in vitro T cell proliferation

Viral-Immunobiology Laboratory, Division of Virology, Department of Neuropharmacology, The Scripps Research Institute, 10666 North Torrey Pines Road, La Jolla, CA 92037, USA

Table 1. Examples of mechanisms by which viruses other than measles can produce a generalized suppression of the host immune response

Mechanism	Examples	References
Virus is directly cytopathic for cells of the immune system	? Human immunodeficiency virus type I	SOMASUNDARAN and ROBINSON 1987; KLATZMANN et al. 1984a,b; DALGLEISH et al. 1984;
Virus infects immune system cells and renders them targets for destruction by antiviral immune effector mechanisms	Lymphocytic choriomeningitis virus ? Human immunodeficiency virus type I	LEIST et al. 1988; BORROW et al. 1991; WALKER et al. 1986;
Virus causes noncytopathic infection of immune system cells but the function of infected cells is impaired	Influenza Human cytomegalovirus ? Epstein-Barr virus (B cells) ? Human immunodeficiency virus type I (dendritic cells)	CASALI et al. 1984; SCHRIER et al. 1986; EVANS and NIEDERMAN 1982; PATTERSON and KNIGHT 1987; MACETONIA et al. 1992;
Free virus or viral components have a direct suppressive effect on cells of immune system	Feline leukemia virus	MATHES et al. 1978; ROJKO and OLSEN 1984.
Virus encodes homologues of cytokine genes or cytokine binding proteins which suppress immune responses	? Epstein-Barr virus ? Vaccinia	HSU et al. 1990; ALCAMI and SMITH 1992; SPRIGGS et al. 1992;
Suppressive factors are produced by host cells (either infected cells, or cells responding to the infection)	? LP-BM5 complex of murine leukemia virus ? Human immunodeficiency virus type I	SHER et al. 1992; SHEARER and Clerici 1992; MATSUYAMA et al. 1991;
Autoimmune responses elicited following virus infection cause immunodysfunction	? Human immunodeficiency virus type I	GOLDING et al. 1989; DELGLEISH et al. 1992.

FORTHAL et al. 1992), and viral RNA has been detected for longer (at least 11 days, or in some studies more than 3 weeks after the onset of the rash (HYYPIA et al. 1985; SCHNEIDER-SCHAULIES et al. 1991).

In vitro, CD4 and CD8- positive T lymphocytes, B lymphocytes and monocytes can all be infected by measles virus, but polymorphonuclear cells cannot (JOSEPH et al. 1975; SULLIVAN et al. 1975b; HUDDLESTONE et al. 1980). Viral replication is restricted in unstimulated PBMCs, and monocytes seem to represent a major infected population in such unstimulated cultures (SALONEN et al. 1980). Measles virus infection of monocytes is not productive even if the cells are activated (VAINIONPAA et al. 1991), although cord blood monocytes from neonates support infection better than monocytes from adult peripheral blood (SULLIVAN et al. 1975b). By contrast, lymphocytes stimulated by a mitogen before or after infection replicate virus, and nearly all cells contain viral antigens (JOSEPH et al. 1975; LUCAS et al. 1978b; HUDDLESTONE et al. 1980). Infected and stimulated $CD4^+$ T lymphocytes were found to support higher titers of viral replication than did $CD8^+$ cells, which correlated inversely with the level of interferon (IFN)-α they produced (JACOBSON and MCFARLAND 1982). The nature of the block to measles virus replication in unstimulated cells and basis of the enhanced permissiveness in activated lymphocytes is not well understood. Cellular proteins have been identified which bind specifically to the 3' noncoding region of the positive strand leader sequence of measles virus RNA and are found only in cells permissive to measles virus infection (LEOPARDI et al. 1993a). Mitogen activation of lymphocytes is associated with an increase in this RNA binding activity, although whether this is the cause of the enhanced permissiveness is not clear (LEOPARDI 1993).

Addressing the question of which PBMC type(s) are infected with measles virus in vivo, an early report found that T and B lymphocytes were among the infected cell populations (WHITTLE et al. 1978). However recent work by ESOLEN et al. (1993) suggests that the majority of infected PBMCs are in fact monocytes. These authors used a measles virus-specific reverse transcriptase-polymerase chain reaction to localize the infected cells to the plastic adherent fraction of PBMCs and combined in situ hybridization for the measles virus fusion protein (F) gene and esterase staining to identify infected PBMCs as monocytes. Esterase-positive cells in the spleen were also found to be infected during acute measles virus infection in vivo (ESOLEN et al. 1993). In the light of what is known of measles virus–PBMC interactions in vitro, whilst the observed infection of monocyte/macrophages would be predicted, it seems surprising that no infection of lymphocytes was found, especially considering that there is a period of pronounced spontaneous proliferation of PBMCs during the first week of the measles rash (GRIFFIN et al. 1986, 1989), at the peak of which 5%–10% of peripheral blood T lymphocytes and more than one third of B lymphocytes are preparing to divide (WARD et al. 1990) and thus should be able to support productive measles virus replication. One possible explanation might be that productively infected PBMCs i.e., lymphocytes but not monocytes, are rapidly removed from the circulation (perhaps via trapping by cells of the reticuloendothelial system or in lymph nodes) and were thus not scored in the patient samples analyzed.

3 Suppression of In Vitro Immune Responses by Measles Virus

A variety of in vitro immune responses are inhibited following infection of PBMCs with measles virus (reviewed by McCHESNEY and OLDSTONE 1987; 1989; CASALI et al. 1989). What is suppressed is the generation of effector functions; already differentiated effector functions are not affected (GALAMA et al. 1980; CASALI et al. 1984). For example, proliferation and generation of cytotoxic T lymphocyte (CTL) activity in allogeneic mixed lymphocyte reaction (MLR) cultures are decreased if virus is added near the start of the culture period, but infection of 5-day-old MLR cultures does not affect the lytic activity of allogeneic CTLs (GALAMA et al. 1980). Similarly if measles virus is added to pokeweed mitogen (PWM)-stimulated PBMC cultures during the first three days, B cell immunoglobulin (Ig) secretion is inhibited, but if the virus is added after day 3, less suppression of Ig secretion is observed (GALAMA et al. 1980; CASALI et al. 1984). As the generation of immune responses, the step which is sensitive to measles virus suppression, requires lymphocyte proliferation, and measles virus has long been known to be able to suppress the proliferation of PBMCs (SMITHWICK and BERKOVICH 1969; ZWEIMAN 1971), McCHESNEY and OLDSONE (1989) have suggested that the multiple effects of measles virus on in vitro lymphocyte functions may all be mediated through a common pathway—the suppression of lymphocyte proliferation.

The mechanism(s) by which measles virus suppresses lymphocyte proliferation have been the subject of much research. Infectious virus is required to produce suppression: inactivation of measles virus by heat, UV light or antibody neutralization (LUCAS et al. 1977; CASALI et al. 1984) or culture at a temperature nonpermissive for viral replication abrogates the suppression. Virus killing of infected lymphocytes has thus been suggested as one possible mechanism for immune suppression. Its importance in the production of immune suppression is measles virus strain-dependent. Whilst productive infection of lymphocytes is not lytic (LUCAS et al. 1977; 1978b; CASALI et al. 1984; McCHESNEY et al. 1986), measles virus infection may lead to syncytia formation and hence cell death (ENDERS and PEEBLES 1954; BARRY et al. 1976). Certain measles virus isolates, e.g., the Halle measles strain, originally isolated from an SSPE patient, cause extensive syncytium formation in PHA-stimulated lymphocytes, leading to cell death within 3 days (ILONEN et al. 1988). Cell death thus may play an important role in the rapid, strong inhibition of PHA-stimulated lymphocyte proliferation produced by these virus strains (ILONEN et al. 1988). By contrast, the Edmonston and certain other strains of measles virus do not produce rapid widespread syncytia formation in infected lymphocyte suspension cultures, and although cell death becomes apparent in extended cultures, inhibition of T cell proliferation following PHA stimulation is observed well before cell viability becomes compromised (ILONEN et al. 1988; SALONEN et al. 1989; YANAGI et al. 1992). A further piece of evidence which demonstrates that measles virus is able to suppress lymphocyte proliferation by a mechanism other than affecting cell viability is the report that a temperature sensitive measles virus mutant, MV ts38, was able to inhibit T cell mitogenesis at 37 °C, a temperature at which it did not undergo a full cycle of replication or induce any cytopathic effect (VYDELINGUM et al. 1989).

Whether measles virus-induced suppression of lymphocyte proliferation is caused by a direct effect of the virus on lymphocytes or by an effect on monocyte/macrophages or other antigen presenting cells has also been a subject of much controversy. A role for monocytes in the mechanism by which measles virus suppresses T lymphocyte proliferation in vitro was suggested by studies of SALONEN et al. (1989) who found that PHA-induced proliferation of lymphocyte-enriched PBMC populations (with <1% contaminating macrophages) was not suppressed at 48 h postinfection, whereas the proliferation of cultures enriched for monocytes (less than 10% of the cells in which were T cells) was more strongly suppressed than that of unfractionated PBMCs. This might be explained by the fact that measles virus infection triggers the production of IFN-α (a cytokine known to have an inhibitory effect on T and B cell proliferation) from monocytes, as the addition of anti-IFN-α antiserum partly reversed the virus-induced inhibition (SALONEN et al. 1989). However other investigators have excluded both IFN-α and prostaglandin E as mediators of the suppression of lymphocyte proliferation by measles virus (LUCAS et al. 1978a: SANCHEZ-LANIER et al. 1988). Recent studies have also shown that measles virus infection of peripheral blood monocytes in vitro causes abnormalities in their functioning: interleukin (IL)-1β expression is enhanced and tumor necrosis factor-α (TNF-α) expression reduced (LEOPARDI et al. 1992), and the infected cells have a reduced capacity to present antigens other than measles virus antigens, despite increases in their level of surface MHC class II expression (LEOPARDI et al. 1993b). The role these changes play in measles virus suppression of lymphocyte proliferation in vitro is currently unclear.

However there is also strong evidence that measles virus can affect lymphocyte functioning directly (SULLIVAN et al. 1975a; LUCAS et al. 1978a; MCCHESNEY et al. 1986). For example, studying measles virus-induced suppression of B cell immunoglobulin production in PWM-stimulated PBMC cultures, MCCHESNEY et al. (1986) showed that whereas purified B cells cultured with PWM in conditioned medium from either measles virus-infected or uninfected PWM-stimulated T cells plus monocyte cultures (cleared of infectious virus by ultracentrifugation) secreted IgG and IgM, infected B cells secreted little immunoglobulin in any conditioned medium, i.e. the virus was directly affecting B cells. In order to determine at which stage in the cell cycle virus-induced suppression was occurring, these authors (MCCHESNEY et al. 1987, 1988) went on to examine events of B cell and T cell activation in measles virus-infected lymphocytes. They found that mitogenic stimulation of infected lymphocytes resulted in cell volume increases, induction of rapid early RNA synthesis and expression of cell surface activation antigens including 4F2, MHC class II, transferrin receptors and (on T cells) IL-2 receptors, as in uninfected cells. The capacity of infected T cells to secrete IL-2 and IFN-γ after stimulation was not reduced, explaining why addition of exogenous IL-2 to measles virus-infected PBMCs does not restore the proliferative response (BORYSIEWICZ et al. 1985). However total RNA synthesis was decreased relative to uninfected cells at 48 and 72 h, and this was correlated with a reduction in the steady-state level of mRNA for histone 2B, one of a group of genes normally upregulated during the S phase of the cell cycle, at 72 h after stimulation. These data suggest that measles virus infection of lymphocytes results in a growth factor-independent arrest in the late G_1 phase of the cell cycle. Further experiments showed that the arrest of progression to the S phase in measles virus-infected T cells is not complete: a low frequency of infected cells do start DNA synthesis, indicating that measles virus infection per se does not completely block DNA synthesis in infected cells (Yanagi et al. 1992).

Thus measles virus infection does not significantly affect the lymphocyte activation process, nor does it cause the complete arrest of DNA synthesis in infected lymphocytes. One hypothesis to explain these observations is the idea of viral interference with the "luxury functions" of differentiated cells (Fig. 1). This hypothesis proposes that although virus infection does not result in host cell lysis or disturbance of housekeeping functions, it interferes with the cell's ability to make product(s) necessary for its differentiated functions—in the case of lymphocytes, product(s) necessary for their immune function. An alternative interpretation of these results, proposed by Yanagi et al. (1992), is that measles virus does

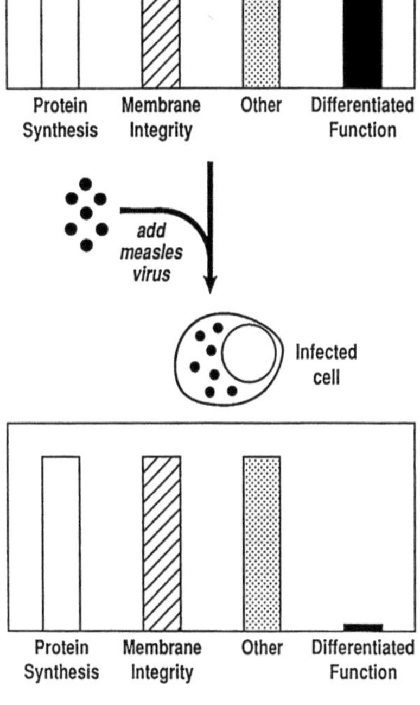

Fig. 1. Virus-induced interference with the differentiated functions of cells: one mechanism which may be involved in measles virus-induced suppression of immune responses. According to this scenario, measles virus replication within lymphocytes and/or antigen presenting cells does not result in cell lysis or interference with housekeeping functions, e.g. protein synthesis is not shut off, but the infection selectively interferes with the cells' ability to make product(s) necessary for their differentiated function i.e. role in the immune response. This concept, and examples of the role it plays in the pathogenesis of a number of different virus infections, is reviewed by Oldstone (1984); de la Torre et al. (1991).

not affect T cell activation and proliferation intracellularly, but exerts its effects through some mechanism related to the expression of its gene products on the cell membrane. The majority of infected lymphocytes express measles virus hemagglutinin (H) and fusion (F) proteins on their surface by 48 h after PHA stimulation: cell-cell interactions mediated through these proteins may initiate intracellular signals that affect cell proliferation (Fig.2). Just as virus binding to cells has been shown to influence specific gene or cell activation events (Gazzolo and Dodon 1987; Boldogh et al. 1990; Peston 1990), the binding of viral gene products to cells may also have a similar effect. Experimental evidence in support of this hypothesis includes the findings: (a) that anti-measles antibodies can prevent the inhibitory effect of measles virus on T cell proliferation even when added 24 h after the initiation of infection (Lucas et al. 1977; Yanagi et al. 1992), (b) that a synthetic oligopeptide Z-D-Phe-L-Phe-Gly which suppresses measles virus-induced fusion, probably by preventing the interaction of measles virus H or F protein with the cell membrane, also reversed the inhibition of T cell proliferation by measles virus (Yanagi et al. 1992), and (c) that measles virus-infected, irradiated PBMCs, when added to fresh autologous PBMCs, caused significant inhibition of T cell proliferation even though the level of virus recovered from the irradiated infected cells was 100-fold lower than the virus titer needed to cause the equivalent inhibition. Furthermore, this inhibition was reversed by anti-measles antibody (Sanchez-Lanier et al. 1988).

This is an attractive hypothesis, because if, as suggested in Fig. 2, the interaction of measles virus proteins on the surface of an infected cell with uninfected lymphocytes is sufficient to inhibit the proliferation of the uninfected

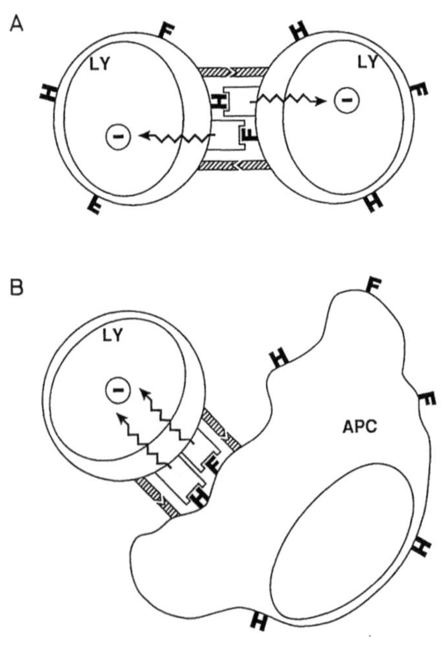

Fig. 2. A second hypothetical mechanism by which measles virus-induced suppression of lymphocyte proliferation may occur: via cell–cell interactions mediated by measles virus glycoproteins. The interaction of a lymphocyte (LY) with measles virus hemagglutinin (H) and/or fusion (F) proteins expressed on the surface of an infected cell delivers a transmembrane signal (zig-zag arrows) to the responding cell which arrests its proliferation (minus sign). Adhesion proteins (striped bars) may facilitate suppression by enhancing cell–cell interactions. **A** Whether such a suppressive cell–cell interaction can occur only between infected lymphocytes, or **B,** whether e.g., the response of an uninfected lymphocyte could be affected following interaction with measles proteins expressed on the surface of an infected antigen presenting cell (APC) is currently unknown

cells (supported by the work of SANCHEZ-LANIER et al. 1988), it would allow measles virus infection of a relatively small number of cells to have a far-reaching effect on the immune response mounted by a large population of uninfected cells. The recent finding that leukocyte function-associated molecule-1 (LFA-1) expression is up-regulated on measles virus-infected leukocytes, promoting cellular aggregation (ATTIBELE et al. 1993) is of interest here, as it might explain why measles virus infected cells may engage in a large number of cell-cell interactions.

4 Mechanisms That May Contribute to the Immune Suppression Associated with Measles Virus Infection In Vivo

Theories as to the mechanism of measles virus-induced immune suppression in vivo must both explain the duration of the immune suppression [cellular immune responses including delayed-type hypersensitivity skin test responses (VON PIRQUET 1908; WESLEY et al. 1978), NK cell activity (GRIFFIN et al. 1990), and in vitro lymphoproliferative responses to mitogens (SMITHWICK and BERKOVICH 1969; WESELY et al. 1978; HIRSCH et al. 1984) are depressed from the onset of the rash for more than 6 weeks], and take into account the findings that measles virus can be isolated from peripheral blood leukocytes only during the early phase of infection and that the proportion of PBMCs which have been shown to be infected is low (OSUNKOYA et al. 1974; WHITTLE et al. 1978; HYYPIA et al. 1985; SCHNEIDER-SCHAULIES et al. 1991; ESOLEN et al. 1993). As with the suppression of immune responses seen following infection of PBMCs with measles virus in vitro, a number of different hypotheses as to the mechanisms involved in the immune suppression associated with natural measles infection have been put forward. It seems likely that a succession of different abnormalities may contribute to the prolonged period of reduced immune responsiveness, rather than a single defect being responsible.

Lymphopenia resulting from destruction of measles virus-infected cells is one of the factors which has been proposed to contribute to immune suppression in vivo. That lymphopenia, which seems to affect B cells and T cells of the $CD4^+$ and $CD8^+$ subsets to a similar extent as there is no change in the ratio of cells of these subsets, occurs during the acute phase of measles has been documented by a number of investigators (WESLEY et al. 1978; WHITTLE et al. 1978; ARNEBORN and BIBERFELD 1983; GRIFFIN et al, 1986). Whether or not this lymphopenia results from destruction of cells infected with measles virus is unclear. A period of spontaneous proliferation of mononuclear cells begins just before the rash appears and continues for several weeks (GRIFFIN et al. 1986; 1989)—during this time a high proportion of PBMCs of all subtypes are activated, and would thus be targets for productive infection by measles virus (WARD et al. 1990). Whilst investigations of the extent of infection of PBMC by measles virus in vivo revealed a low proportion of infected cells, and one report has suggested that the infection may be restricted primarily to monocytes (ESOLEN et al. 1993), it is possible (as discussed earlier) that measles infection of PBMCs is in fact much more extensive, but that infected cells are rapidly removed from the circulation (altered trafficking patterns have been reported for lymphocytes infected with other viruses by WOODRUFF and WOODRUFF 1974); or that, as has been found to be the case in HIV-1 infection in humans (PANTALEO et al. 1993; EMBRETSON et al. 1993), there is a much higher level of infection of mononuclear cells in lymph nodes and other lymphoid organs than is reflected in circulating PBMCs. Whatever the cause of the observed lymphopenia, however, it alone cannot entirely account for the clinical immune suppression associated with measles virus infection, as depression of immune function persists beyond the lymphopenic phase of the disease, suggesting that abnormalities in cellular functioning also occur.

The role of abnormalities in monocyte functioning in the depression of cellular immune responses during natural measles infection was addressed by GRIFFIN et al. (1987) and WARD et al. (1991). They found a variety of abnormalities in monocyte function in measles patients: in particular monocyte production of TNF-α, a cytokine which synergizes with other cytokines early during immune responses to support T and B cell proliferation, was frequently reduced during the first week after onset of the measles virus rash. However, overall these data suggested that abnormal monocyte function was probably not the primary cause of measles virus-associated immunosuppression.

This implies that, as in vitro, the in vivo immune suppression associated with measles virus infection may be mediated at the level of lymphocyte responsiveness. Whether or not the mechanisms proposed to explain measles virus-induced suppression of lymphocyte proliferation in vitro, namely that either infection internally disturbs the differentiated functions of lymphocytes, or alternatively that interaction of the responding lymphocytes with measles virus H and/or F proteins on the surface of infected cells delivers a transmembrane signal to the responding cells which arrests their proliferation, also operate in vivo is unknown. However the latter type of mechanism would provide an explanation for why profound suppression of lymphocyte responses occurs despite the fact that virus infection is detected only in a small proportion of monocytes (ESOLEN et al. 1993): each infected monocyte may interact with a succession of lymphocytes and thus render a large number of the latter cells anergic.

Considering immune suppression in vivo, other mechanisms involving evolution of the immune response which are less likely to play an important role in the suppression of lymphocyte functioning occurring 48 h after in vitro infection with measles virus are also possible. Evidence for one such mechanism in the immune suppression accompanying natural measles virus infection comes from studies of the type of immune response mounted following measles infection (reviewed in greater detail elsewhere in this volume). Briefly, examination of T cell cytokine production, both by measurement of plasma levels of different cytokines during natural measles infection and by assessing in vitro cytokine production in response to stimulation with mitogens or an anti-CD3 antibody, from T cells of measles patients (WARD et al. 1991; GRIFFIN and WARD 1993), revealed that during and shortly after the rash, in vitro T cell proliferation could be improved by supplementation with IL-2, although in vivo plasma IL-2 levels were high. After resolution of the rash, plasma IL-4 became elevated, and remained high in some patients for more than 7 weeks. In vitro, mononuclear cells proliferated poorly in response to stimulation with anti-CD3 and produced low levels of IL-2 and IFN-γ and high levels of IL-4 and IL-6, a lymphokine profile typical of a response mounted predominantly by type 2 $CD4^+$ T cells (SALGAME et al. 1991; KAPSENBERG et al. 1991). IL-4 inhibits IL-2-dependent T cell proliferation, and IL-4 production likely contributed to the depressed in vitro lymphoproliferative responses to mitogens as addition of antibody to IL-4 partially corrected the proliferative defect. However, other factors undoubtedly affect in vitro lymphoproliferation during measles since responses were also low during the rash, when plasma IL-4 levels were normal and anti-IL-4 antibody did not have any effect (GRIFFIN and WARD 1993). In vivo, many of the other long-term immunologic abnormalities associated with measles infection, including decreased macrophage activation, suppression of delayed-type hypersensitivity and low NK cell activity, may be related to preferential activation of type 2 $CD4^+$ T cells.

Thus it is possible that an evolution of immunologic abnormalities affecting both lymphocytes and monocytes occurs over the course of measles-associated immunosuppression. Early in the infection, direct viral effects on lymphocytes and monocyte abnormalities may contribute to immunosuppression, whereas the long-term immune dysfunction occurring after viral clearance likely results from the lymphocyte response becoming programmed into a particular immune profile during which the majority of cellular responses are strongly down-regulated, allowing secondary infections to occur.

5 Conclusions

Natural infection with measles virus is associated with a clinically relevant immune suppression. In an attempt to understand the mechanism(s) by which this is produced, measles virus- mononuclear cell interactions following both in vitro and in vivo contact of cells with virus have been studied.

In vitro, measles virus can infect T and B lymphocytes and monocytes, although the infection is productive only in activated lymphocytes. The generation of effector functions by diverse lymphocyte subpopulations is inhibited following infection, but already differentiated functions are not affected. Inhibition appears to result from the suppression of lymphocyte proliferation; and a number of different mechanisms have been proposed to explain how this is achieved. These include: (a) virus killing of infected lymphocytes, which whilst it may play an important role in the suppression produced by certain highly cytopathic measles virus isolates, cannot fully explain the inhibition of lymphocyte proliferation produced by many other less cytopathic measles virus strains; (b) virus effects on monocytes— again, although measles virus infection of peripheral blood monocytes does cause abnormalities in their functioning, their role in the observed suppression of lymphocyte proliferation is currently unclear; and (c) a direct viral effect on lymphocyte functioning, which may either occur as a consequence of virus internally altering the function of infected cells, or may involve delivery of a negative transmembrane signal to responding lymphocytes following interaction with measles virus H and/or F proteins on the surface of an infected cell. It is possible that several or all of these mechanisms may contribute to the observed immune suppression, their relative importance depending on the virus isolate used, PBMCs subtypes present, and immune function studied.

Measles virus-mononuclear cell interactions in vivo and the mechanism(s) which contribute to the immune suppression in man are even less clearly understood. Infection of a small number of PBMCs, the majority of which appear to be monocytes, can be detected transiently following the onset of the measles virus rash; although whether circulating PBMCs reflect the extent and duration of mononuclear cell infection in the entire immune system is unknown. Here again, a plethora of mechanisms have been suggested to play a role in the production of the observed immune suppression; what seems most likely is that a series of immunologic abnormalities contribute sequentially to immunosuppression. The suppression observed shortly after the onset of the rash may be due to similar mechanisms to those discussed above for the acute suppression of responses following infection of PBMCs with measles virus in vitro, i.e., direct viral effects on lymphocytes and monocyte abnormalities may be the most important contributors. By contrast, the long-term immune dysfunction occurring after viral clearance may result from the induction of a strong type 2 $CD4^+$ T cell immune response during which the majority of cellular responses are down-regulated to a clinically important extent.

Recent observations suggest that approaches to solve the riddle of how measles virus causes immunosuppression in vivo are almost on hand. The CD46 molecule has been identified as a cellular receptor for measles virus (Naniche et al. 1993). This observation has been confirmed and it has been documented that transfection of various isoforms of CD46 into nonpermissive hamster or murine cells renders them permissive for measles virus infection (Manchester et al. 1994). Transgenic mice bearing the CD46 gene have thus been constructed (Manchester, Atkinson and Oldstone, unpublished data) to serve as a small animal model for the study of measles virus infection and the associated immunosuppression. Similarly measles virus has been shown to induce immunosuppression in vivo in a severe combined immunodeficiency (SCID)-human PBMC mouse model (Manchester, Tishon and Oldstone, unpublished data). The ability to manipulate the immune response by depletion and reconstitution in such animal models offers the possibility of solving the puzzle of measles virus-induced immunosuppression first reported by Clement von Pirquet nearly 100 years ago.

Acknowledgements. This is publication number 8283-NP from the Department of Neuropharmacology, the Scripps Research Institute. This work was supported by NIH grants NS12428 and AI09484 and by training grant MH47680. The authors would like to thank Kim Fishman for typing this manuscript.

References

Alcami A, Smith GL (1992) A soluble receptor for interleukin-1β encoded by vaccinia virus: a novel mechanism of virus modulation of the host response to infection Cell 71: 153–167

Arneborn P, Biberfeld G (1983) T-lymphocyte subpopulations in relation to immunosuppression in measles and varicella Infect Immun 39: 29–37

Attibele N, Wyde PR, Trial J, Smole SC, Smith CW, Rossen RD (1993) Measles virus-induced changes in leukocyte function antigen 1 expression and leukocyte aggregation: possible role in measles virus pathogenesis J Virol 67: 1075–1079

Barry DW, Sullivan JL, Lucas SJ, Dunlap RC, Albrecht P (1976) Acute and chronic infection of human lymphoblastoid cell lines with measles virus, J Immunol 116: 89–98

Beckford AP, Kaschula ROC, Stephen C (1985) Factors associated with fatal cases of measles: a retrospective autopsy study, S Afr Med J 68: 858–863

Blumberg RW, Cassady HA (1947) Effect of measles on nephrotic syndrome Am J Dis Child 63: 151–166

Boldogh I, Abubaker S, Albrecht T (1990) Activation of protooncogenes: an immediate early event in human cytomegalovirus infection. Science 247: 561–564

Borrow P, Tishon A, Oldstone MBA (1991) Infection of lymphocytes by a virus that aborts cytotoxic T lymphocyte activity and establishes persistent infection. J Exp Med 174: 203–212

Borysiewicz LK, Casali P, Rogers B, Morris S, Sissons JGP (1985). The immunosuppressive effects of measles virus on T cell function-failure to affect IL-2 release or cytotoxic T cell activity in vitro. Clin Exp Immunol 59: 29–36

Casali P, Rice GP, Oldstone MBA (1984) Viruses distrupt functions of human lymphocytes. Effects of measles virus and influenza virus on lymphocyte-mediated killing and antibody production. J Exp Med 159: 1322–1337

Casali P, Nakamura M, McChesney MB (1989) Immunosuppression by measles virus. In: Spector S, Bendinelli M, Friendman H (eds) Virus-induced immunosuppression. Plenum, New York, pp 345–373

Dalgleish AC, Beverley PCL, Clapham PR, Crawford DH, Greaves MF, Weiss RA (1984) The CD4(T4) antigen is an essential component of the receptor for the AIDS retrovirus. Nature 312: 763–767

Dalgleish AG, Wilson S, Gompels M, Ludlam C, Gazzard B, Coates AM, Habeshaw J (1992) T-cell receptor variable gene products and early HIV-1 infection, Lancet 339: 824–828

de la Torre JC, Borrow P, Oldstone MBA (1991) Viral persistence and disease: cytopathology in the absence of cytolysis Br Med Bull 47: 838–851

Embretson J, Zupancic M, Ribas JL, Burke A, Racz P, Tenner-Racz K, Haase AT (1993) Massive covert infection of helper T lymphocytes and macrophages by HIV during the incubation period of AIDS, Nature 362: 359–362

Enders JF, Peebles TC (1954) Propagation in tissue cultures of cytopathogenic agents from patients with measles, Proc Soc Exp Biol Med 86: 277–286

Esolen LM, Ward BJ, Moench TR, Griffin DE (1993) Infection of monocytes during measles. J Infect Dis 168: 47–52

Evans AS, Niederman JC (1982) In viral infections of humans: epidemiology and control Plenum, New York, pp253–281

Fireman P, Friday G, Kumate J (1969) Effect of measles vaccine on immunologic responsiveness. Pediatrics 43: 264–272

Forthal DN, Aarnaes S, Blanding J, de la Maza L, Tilles JG (1992) Degree and length of viremia in adults with measles. J Infect Dis 166: 421–424

Galama JMD, Ubels-Postma J, Vos A, Lucas CJ (1980) Measles virus inhibits acquisition of lymphocyte functions but not established effect or functions. Cell Immunol 50: 405–415

Gazzolo L, Dodon MD (1987) Direct activation of resting T lymphocytes by human T-lymphotropic virus type I. Nature 326: 714–717

Golding H, Shever GM, Hillman K, Lucas P, Manischewitz J, Zajac RA, Clerici M, Gress RE, Boswell BN, Golding B (1989) Common epitope in human immunodeficiency virus (HIV)-1 gp41 and HLA class II elicits suppressive autoantibodies capable of contributing to immune dysfunction in HIV-1-infected individuals. J Clin Invest 83: 1430–1435

Gresser J, Chany C (1963) Isolating of measles virus from the washed leucocytic fraction of blood. Proc Soc Exp Biol Med 113: 695–698

Griffin DE, Ward BJ (1993) Differential CD4 T cell activation in measles. J Infect Dis 168: 275–281

Griffin DE, Moench TR, Johnson RT, de Soriano IL Vaisberg A (1986) Peripheral blood mononuclear cells during natural measles virus infection: cell surface phenotypes & evidence for activation. Cline Immunol Immunopathol 40: 305–312
Griffin DE, Johnson RT, Tamashiro VG, Moench TR, Jauregui E, Lindo-de-Soriano I, Vaisberg A (1987) In vitro studies of the role of monocytes in the immunosuppression asssociated with natural measles virus infections. Cline Immunol Immunopathol 45: 375–383
Griffin DE, Ward BJ, Johnson RT, Jauregui E, Vaisberg A (1989) Immune activation in measles. N Engl J Med 320: 1667–1672
Griffin DE, Ward BJ, Jauregui E, Johnson RT, Vaisberg A (1990) Natural killer cell activity during measles. Clin Exp Immunol 81: 218–224
Hirsch HL, Griffin DE, Johnson RT, Cooper SJ, Lindo de Soriano I, Roedenbeck S, Vaisberg A (1984) Cellular immune responses during complicated and uncomplicated measles virus infections of man, Clin Immunol Immunopathol 31: 1–12
Hsu DH, de Waal Malefyt R, Fiorentino DF, Dang MN, Vieira P, de Vries J, Spits H, Mosmann RT, Moore KW (1990) Expression of interleukin-10 activity by Epstein-Barr virus protein BCRF1, Science 250: 830–832
Huddlestone JR, Lampert PW, Oldstone MBA (1980) Virus-lymphocyte interactions: infection of T_G and T_M subsets by measles virus. Cline Immunol Immunopathol 15: 502–509
Hyypia T, Korkiamaki P, Vainionpaa R (1985) Replication of measles virus in human lymphocytes. J Exp Med 161: 1261–1271
Ilonen I, Salonen R, Marusyk R, Salmi A (1988) Measles virus strain-dependent variation in outcome of infection of human blood mononuclear cells. J. Gen Virol 69: 247–254
Jacobson S, McFarland HF (1982) Measles virus persistence in human lymphocytes: a role for virus-induced interferon J Gen Virol 63: 351–357
Joseph BS, Lampert PW, Oldstone MBA (1975) Replication and persistence of measles virus in defined subpopulations of human leukocytes. J Virol 16: 1638–1649
Kapsenberg ML, Wierenga EA, Bos JD, Jansen HM (1991) Functional subsets of allergenreactive human CD4+ T cells. Immunol Today 12: 392–395
Klatzmann D, Champagne E, Chamaret S, Gruest J, Guetard D, Hercend T, Gluckman JC Montagnier L (1984b) T-Lymphocyte T4 molecule behaves as the receptor for human retrovirus LAV. Nature 312: 767–768
Leist TP, Ruedi E, Zinkernagerl EM (1988) Virus-triggered immune supperession in mice caused by virus-specific cytotoxic T cells. J Exp Med 167: 1749–1754
Leopardi R (1993) Measles virus infection in human immune cells Turun Yliopiston Julkaisuja Annales Universitatis Turkuensis Sarja-Ser D. Med Odontol 114: 1–29
Leopardi R, Vainionpaa R, Hurme M, Siljander P, Salmi A (1992) Measles virus infection enhances IL-1b but reduces tumor necrosis factor-α expression human monocytes. J Immunol 149: 2397–2401
Leopardi R, Hukkanen V, Vainionpaa R, Salmi A (1993a) Cell proteins bind to sites within the 3' noncoding region and the positive-strand leader sequence of measles virus RNA. J Virol 67: 785–790
Leopardi R, Ilonen J, Mattila L, Salmi A (1993b) Effect of measles virus infection on MHC class II expression and antigen presentation in human monocytes. Cell Immunol 147: 388–396
Lucas CJ, Galana JMD, Ubels-Postma J (1977) Measles virus - induced suppression of lymphocyte reactivity in vitro. Cell Immunol 32: 70–85
Lucas CJ, Ubels-Postma JC, Galana JMD, Rezee A (1978a) Studies on the mechanism of measles virus-induced suppression of lymphocyte functions in vitro. Cell Immunol 37: 448–458
Lucas CJ, Ubels-Postma JC, Rezee A Galana JMD, Rezee A (1978b) Activation of measles virus from silently-infected human lymphocytes. J Exp Med 148: 940–952
Macetonia SE, Gompels M, Pinching AJ, Patterson S, Knight SC (1992) Antigen presentation by macrophages but not by dendritic cells in human immunodeficiency virus (HIV) infection. Immunology 75: 576–581
Manchester M, Liszewski MK, Atkinson J, Oldstone MBA (1993) Multiple isoforms of CD46 (membrane cofactor protein) can serve as receptors for measles virus. Proc Natl Acad Sci USA 91: 2161–2165
Mathes LE, Olsen RC, Habebrand LC, Hoover EA, Schaller JP (1978) Abrogation of lymphocyte blastogenesis by a feline leukemia virus protein. Nature 274: 687–689
Matsuyama T, Kobayashi N, Yamamoto N (1991) Cytokine and HIV infection: is AIDS a TNF disease? AIDS 5: 1405–1417
McChesney MB, Oldstone MBA (1987) Viruses perturb lymphocyte functions: selected principles characterizing virus-induced immunosuppression. Annu Rev Immunol 5: 279–304
McChesney MB, Oldstone MBA (1989) Viruses - induced immunosuppression: infections with measles virus and human immunodeficiency virus. Adv Immunol 45: 335–380

McChesney MB, Fujinami RS, Lampert PW, Oldstone MBA (1986) Viruses disrupt functions of human lymphocytes II. Measles virus suppresses antibody production by acting on B lymphocytes. J Exp Med 163: 1331–1336

McChesney MB, Kehrl JH, Valsamakis A, Fauci AS, Oldstone MBA (1987) Measles virus infection of B lymphocytes permits cellular activation but block progression through the cell cycle. J Virol 61: 3441–3447

McChesney MB, Altman A, Oldstone MBA (1988) Suppression of T lymphocyte function by measles virus is due to cell cycle arrest in G1 J Immunol 140: 1269–1273

Miller DL (1964) Frequency of complication of measles 1963 Br Med J 2: 75–78

Naniche D, Varior-Krishnan G, Cervoni F, Fabian Wild T Rossi B, Rabourdin-Combe C, Gerlier D (1993) Human membrane cofactor protein (CD46) acts as a cellular receptor for measles virus. J Virol 67: 6025–6032

Oldstone MBA (1984) Virus can alter cell function without causing cell pathology: discordered function leads to imbalance of homeostasis and disesase. In: Notkins AL, Oldstone MBA Concepts in viral pathogenesis Springer, Berlin Heidelberg New York, pp 269–276

Osunkoya BO, Adeleye GI, Adejumo TA, Salimonu LS (1974) Studies on leukocyte cultures in measles II. Detection of measles virus antigen in human leukocytes by immunofluoresence. Arch Gesamte Virusforsch 44: 323–329

Pantaleo G, Graziosi C, Demarest JF, Butini L, Montroni M, Fox CH, Orenstein JM, Kotler DP, Fauci AS (1993) HIV infection is active and progressive in lymphoid tissue during the clinically latent stage of disease. Nature 362: 355–358

Patterson S, Knight S (1987) Susceptibility of human peripheral blood dendritic cells to infection by human immunodeficiency virus. J Gen Virol 68: 1177–1181

Papp K (1937) Fixation de virus morbilleux aux leucocytes du sang de la periode d'incubation de la maladie. Bull Acad Med (Paris) 117: 46–51

Peebles TC (1967) Distribution of virus in blood components during the viremia of measles. Arch Gesamte Virus forsch 22: 43–47

Preston VG (1990) Hepes simplex virus activates expression of a cellular gene by specific binding to the cell surface. Virology 176: 474–482

Rojko RL, Olsen RG (1984) The immunobiology of the feline leukemia virus. Vet Immunol Immunopathol 6: 107–165

Salonen R, Ilonen J. Salmi AA (1988) Measles virus infection of unstimulated blood mononuclear cells in vitro: antigen expression and virus production preferentially in monocytes. Clin Exp Immunol 71: 224–228

Salonen R, Ilonen J. Salmi AA (1989) Measles virus inhibits lymphocyte proliferation in vitro by two different mechanisms. Clin Exp Immunol 75: 376–380

Salgame P, Abrams JS, Clayberger C, Goldstein H, Convit J, Modlin RL, Bloom BR (1991) Differeing lymphokine profiles of functional subsets of human CD4 and CD8 T cell clones. Scinece 254: 279–282

Sanchez-Lanier M, Guerin P, McLaren LC, Bankhurst AD (1988) Measles virus-induced suppression of lymphocyte proliferation. Cell Immunol 116: 367–381

Schneider-Schaulies S, Kreth HW, Hofman G, Billeter M, ter Meulen V (1991) Expression of measles virus RNA in peripheral blood mononucler cells of patients with measles, SSPE and autoimmune diseases. Virology 182: 703–711

Schrier RD, Rice GPA, Oldstone MBA (1986) Suppression of natural killer cell activity and T cell proliferation by fresh isolates of human cytomegalovirus. J Infect Dis 153: 1084–1091

Shearer GM, Clerici M (1992) T helper cell immune dysfunction in asymptomatic, HIV-1 seropositive individuals: the role of TH1-TH2 cross regulation. In RL Coffman (ed) Regulation and functional significance of T-cell subsets. Chem Immunol 54: 21–43

Sher A, Gazzinelli RT, Oswald IP, Clerici M, Kullberg M, Pearce EJ, Berzofsky JA, Mosmann TR, James SL, Morse HC (1992) Role of T-cell derived cytokinnes in the downregulation of immune responses in parasitic and retroviral infection. Immunol Rev 127: 283–294

Smithwick EM, Berkovich S (1969) The effect of measles on the In vitro lymphocyte response to tuberculin. In: Smith RT, Good RA Cellular recognition. Appleton Centruey Crafts, New York, p 131

Somasundaran M, Robinson HL (1987) A major mechanism of human immunodeficiency virus-induced cell killing does not invove cell fusion. J Virol 61: 3114–3119

Spriggs MK, Hruby DE, Maliszewski CR, Pickup DJ, Sims JE, Buller RML, Vanslyke J (1992) Vaccinia and cowpox viruses encode a novel secreted interleukin-1-binding protein. Cell 71: 145–152

Sullivan JL, Barry DW, Lucas SJ, Albrecht P (1975b) Measles infection of human mononuclear cells. I. Acute infection of peripheral blood lymphocytes and monocytes. J Exp Med 142: 773–784

Vainionpaa R, Hyypia T, Akerman KEO (1991) Early signal transduction in measles virus-infected lymphocytes is unaltered, but second meassangers activate virus replication. J Virol 65: 6743–6748
Von Pirquet (1908) Das verhalten der kutanen tuberculinreaktion wahrend der masern. Dtsch Med Wochenschr 30: 1297–1300
Vydelingum S, Ilonen J, Salonen R, Marusyk R, Salmi A (1989) Infection of human peripheral blood mononuclerar cells with a temperature-sensitive mutant of measles virus. J Virol 63: 689–695
Walker CJ, Mody DJ, Stites DP, Levy JA (1986) CD8+ lymphocytes can control HIV infection in vitro by suppressing virus replication. Science 234: 1563–1566
Ward BJ, Johnson RT, Vaisberg A, Jauregui E, Griffin DE (1990) Spontaneous proliferation of peripheral mononuclear cells in natural measles virus infection: identification of dividing cells and correlation with mitogen reponsiveness. Clin Immunol Immunopathol 55: 315–326
Ward BJ, Johnson RT, Vaisberg A, Jauregui E, Griffin DE (1991) Cytokine production in vitro and the lymphoproliferative defect of natural measles virus infection. Clin Immunol Immunopathol 61: 236–248
Weiss RA (1993) How does HIV cause AIDS? Science 260: 1273–1279
Wesley A, Coovadia HM, Henderson L (1978) Immunological recovery after measles. Clin Exp Immunol 32: 540–544
White RG, Boyud JF (1973) The effect of measles on the thymus and other lymphoid tissues. Clin Exp Immunol 13: 323–357
Whittle HC, Dossetor J, Oduloju A, Bryceson A, Greenwood BM (1978) Cell mediated immunity during natural measles infection. J Clin Invest 62: 678–684
Woodruff JJ, Woodruff JF (1974) Virus-induced alterations of lymphoid tissues. IV. The effect of Newcastle disease virus on the fate of transfused thoracic duct lymphocytes. Cell Immunol 10: 78–85
Yamanouchi K, Chino F, Kobune F, Kodama H, Tsuruhara T (1973) Growth of measles virus in the lymphoid tissues of monkeys. J Infect Dis 128: 795–799
Yanagi Y, Cubitt BA, Oldstone MBA (1992) Measles virus inhibits mitogen-induced T cell prolifereation but does not directly perturb the T cell activation process inside the cell. Virology 187: 280–289
Zweiman B (1971) In vitro effects of measles virus on proliferating human lymphocytes. J Immunol 106: 1154–1158

Measles Virus Gene Expression in Neural Cells

S. SCHNEIDER-SCHAULIES, J. SCHNEIDER-SCHAULIES, L.M. DUNSTER, and V. TER MEULEN

1 Introduction

The interaction of measles virus (MV) with the human central nervous system (CNS) can lead to acute or chronic disease processes which are of clinical importance. In a fraction of patients, an acute measles encephalitis (APME) appears during or after acute measles probably as a result of virus-induced autoimmune reactions. As rare late complications of measles, subacute sclerosing panencephalitis (SSPE) and measles inclusion body encephalitis (MIBE) develop on the basis of a persistent infection in neural cells (for details see Katz, this volume). These disorders have stimulated numerous studies which have contributed to the understanding of alterations of viral gene expression and

Institute for Virology and Immunobiology, University of Würzburg, Versbacher Strasse 7, 97078 Würzburg, Germany

cellular gene functions in brain tissue (for review see BILLETER and CATTANEO 1991; SCHNEIDER-SCHAULIES and TER MEULEN 1992). Based on these investigations, it is obvious that primary interaction of MV with brain tissue is followed by a nonlytic persistent infection instituted by attenuation of viral gene functions. As convincing evidence has not been mounted for the existence of particular defective viruses initially infecting the CNS (CATTANEO et al. 1989; BACZKO et al. 1993), factors intrinsic or induced in neural cells have been proposed to attenuate primary viral gene expression (KRAUS et al. 1992; SCHNEIDER-SCHAULIES et al. 1993 a,b). In this review we will give a brief overview on how defective and quantitatively altered viral gene functions encountered in persistent MV infections of brain cells may relate to cellular factors and regulatory mechanisms and how cellular gene expression in neural cells is impaired upon the interaction with MV.

2 Alterations of Measles Virus Genomes and Gene Functions in Presistent Brain Infections

Alterations of MV gene expression have mainly been characterized by directly using autopsy brain material of patients with SSPE or cell lines infected with MV isolates from brain tissue. As summarized in Fig. 1, expression of the viral envelope proteins M, F and H has been found to be generally low or even absent in persistent brain infections whereas the integrity of the replicative complex, as indicated by the expression of N and P proteins, is apparently maintained. The altered expression of the envelope proteins has been ascribed to a variety of independent mechanisms. A low abundancy of the corresponding mRNA transcripts (CATTANEO et al. 1987a, b), as reflected in the formation of particularly steep transcription gradients, is accompanied by the alterations of the coding sequences that lead, in extreme cases, to the truncation or complete abolishment of the corresponding reading frames (BILLETER and CATTANEO 1991). The natural error frequency of the viral polymerase and the activity of a cellular double-stranded (ds) RNA-dependent unwinding/modifying enzyme have been proposed to cause the introduction of single point or hypermutations, respectively (Fig. 1) Detailed informations on mutationally affected viral gene functions in persistent brain infections are reviewed by Rima et al. (this volume).

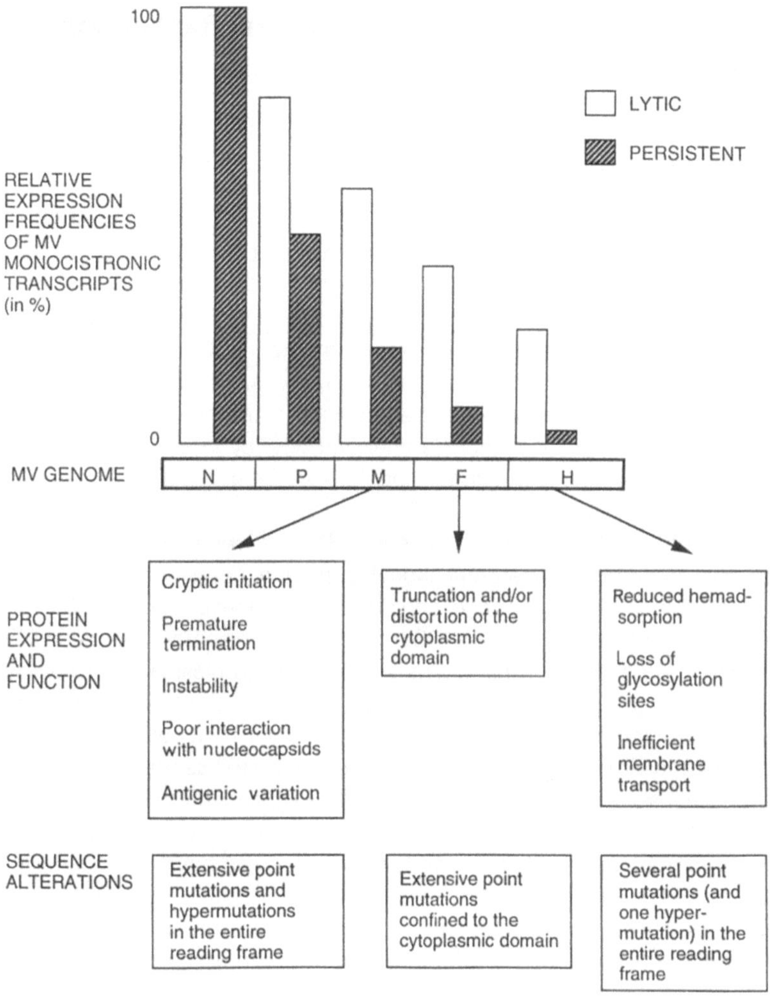

Fig. 1. Alterations of Measles virus (MV) genomes and gene functions in persistent brain infections. Characteristic restrictions of MV gene expression, their basis and functional consequences are summarized. The genome is indicated in the *center* of the diagram in 3' to 5' direction containing all MV structural genes except the L protein since not enough data are available for the particular reading frame. On *top* of the genome, the gradients for the relative frequencies for the monocistronic polyadenylated mRNAs compared to the N gene (N is set to 100%) are indicated. The shallow gradient for the mRNAs expressed in nonneural cells is represented by the *stippled*, the steep gradient for the transcripts present in persistent brain infections by the *hatched boxes*. It should be stressed that the values are chosen arbitrarily as a representative example and do not refer to particular original data. As no pronounced defects for the core proteins N and P have been found, alterations of protein expression are indicated for the envelope proteins M, F and H. In addition to the low frequencies of the corresponding mRNAs the gene products are more or less affected or truncated due to single or clustered nucleotide exchanges

3 Host Cell-Dependent Restrictions of Viral Replication in Brain Cells

3.1 The Experimental Systems

The observation of an impairment of MV gene expression in brain tissue of SSPE patients at the time of death raises the question about the events at primary infection of the CNS. Obviously, as a prerequisite for the establishment of a persistent rather than a lytic infection in brain cells, viral gene functions have to be attenuated in order to avoid the immediate destruction of the host cells. This particular virus-host interaction may be governed by the action of host cell factors imposing certain constraints on viral gene expression, thus favouring viral persistence. Since these questions cannot be analyzed directly in the human brain, experimental systems including animal models have been established (LIEBERT and TER MEULEN 1987; RAMMOHAN et al. 1981; SWOVELAND and JOHNSON 1989).

In an animal model using inbred Lewis and BN rats (see chapter by Liebert and Finke, this volume) intracerebral inoculation of a neurotropic, rat brain-adapted CAM strain of MV induces age- and strain-dependent acute and subacute CNS diseases (LIEBERT and TER MEULEN 1987). These experimentally induced CNS diseases share some of the characteristic neuropathological and virological findings with human persistent brain infections. In contrast to the acute encephalitis, the persistent infection is characterized by (a) a significant humoral immune reaction against the virus, (b) absence of free infectious virus as well as giant cell formation and (c) accumulation of intracellular nucleocapsids (LIEBERT and TER MEULEN 1987; DÖRRIES et al. 1988). In addition to animal models, tissue culture systems with primary and permanent cells of neural origin have been used for descriptive studies to verify the findings in brain tissue (SCHNEIDER-SCHAULIES et al. 1990, 1992). They allowed the investigation of MV gene expression in different brain cell populations dependent on intrinsic factors or in response to external stimuli (SCHNEIDER-SCHAULIES et al. 1990, 1993a; YOSHIKAWA and YAMANOUCHI 1984). Moreover, cell-free extracts have been used to analyze MV transcription and replication in vitro (MOYER et al. 1990; HORIKAMI and MOYER 1991), as well as direct binding studies of cellular transactivating proteins to the regulatory sequences of MV RNA (LEOPARDI et al. 1993). The biological relevance of these findings has still to be confirmed, since in vivo the viral RNA has to be encapsidated by the N protein to serve as a template for the viral polymerase complex (BANERJEE 1987). Although the latter approaches provide some evidence for the requirement of cellular proteins for efficient viral transcription, they have to be evaluated also in view of the fact that cellular structures and compartmentalization are not retained.

3.2 Regulation of Viral Transcription

3.2.1 Intrinsic Factors

A restriction of transcription is common to all persistent MV infections in brain cells (Fig. 1). In analyzing potential host factors it became obvious that MV transcription in cells of neural origin is generally reduced compared to cell systems of nonneural origin, as indicated by the copy numbers of N-specific transcripts per infected cell (SCHNEIDER-SCHAULIES et al. 1989, 1990, 1993a) (Fig. 2). This spontaneous reduction was shown to be up to tenfold in human neural tissue culture cells compared to nonneural controls (SCHNEIDER-SCHAULIES et al. 1993a). In addition, the differentiation state of the cells may interfere with active transcription of MV, as seen in brain material of experimentally infected animals (LIEBERT et al. 1990) and in tissue culture upon in vitro differentiation of neuronal cells (YOSHIKAWA and YAMANOUCHI 1984). As to the overall reduction observed, the

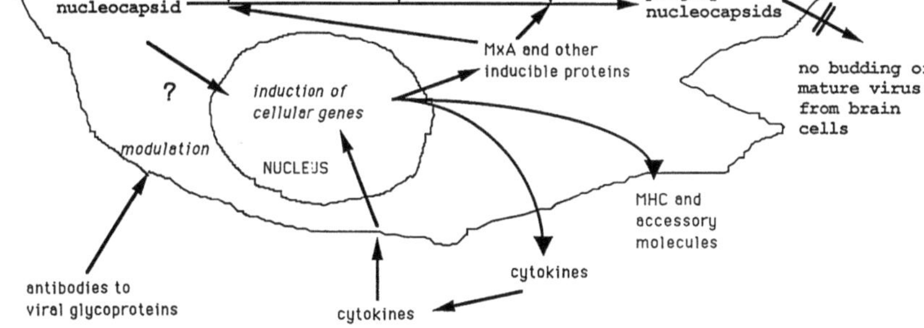

Fig. 2. Host factors influencing Measles Virus (MV) gene expression in neural cells. Host factors and their mode of action contributing to the establishment of persistent CNS infections by modulating viral and cellular gene functions are summarized here. For the infecting virus, total and relative transcriptional efficiencies of its individual mRNAs are downregulated by cellular components present in the cell or induced by cytokines (as MxA) or antiviral antibodies. Although rarely occurring, MV-specific RNAs may be occasionally modified by the action of the cellular unwindase present in the cytoplasm of brain cells. Viral mRNAs, either modified or unmodified, are barely translated into structural proteins, in particular in differentiated brain cells, ending up in a prolonged, inefficient replication of the viral genome and a low expression of the envelope proteins. As a consequence, budding of infectious virus is delayed or may even be absent. As a second event, cellular gene expression is altered upon MV infection, as indicated by the induction of surface molecule expression and the release of a characteristic set of cytokines. To close the circle, these cytokines may enhance not only the antiviral immune response but also directly the intracellular control of viral gene expression as shown for the type I interferon-induced MxA protein

formation of particularly steep transcription gradients, as documented in SSPE brain material (see above), proved to depend to a major extent on cellular control mechanisms. The progressive decrease of transcriptional efficiency along the viral genome leading to a high underrepresentation of the mRNAs for the envelope proteins and, in particular, for the L protein, could be observed in brain tissue of experimentally infected rats and in tissue culture systems with primary and permanent neural cells (SCHNEIDER-SCHAULIES et al. 1989, 1990, 1993a) (Fig. 2.)

Due to the lack of adequate experimental systems, regulation of MV transcription has not been studied intensely yet. At least in vitro, cellular cytoskeletal components are required for activating viral transcription for related viral systems such as human parainfluenza virus type 3 (HPIV3), Sendai and respiratory syncytial virus (RSV) (MOYER et al. 1986; BANERJEE 1987; BARIK 1992; DE et al. 1993). In contrast to the latter viruses, MV transcription seems to require tubulin rather than actin (MOYER et al. 1990). Since these proteins are ubiquitous, they may not contribute to the particular regulation of MV expression in brain cells. More recently, the direct binding of host-encoded proteins to the leader RNA of MV has been suggested to be mandatory for the activation of viral transcription and replication. These host factors have been proposed to be lacking in cells of lower permissivity for MV (LEOPARDI et al. 1993).

As a consequence of the low abundance of the corresponding transcripts, the levels of L protein are most probably rather low in infected brain cells and may account for a limited extent of virus-specific transcription. A certain stoichiometry of N, P and L proteins for the formation of functional complexes has been shown to be required for the related Sendai virus system (HORIKAMI et al. 1992). In addition, an inhibitory capacity of the Sendai virus V protein on viral RNA replication was found in vitro when the L protein was limiting. Thus, V protein may exert a particularly strong negative effect on replication in infections in which the concentration of L is relatively low, such as in brain cells (CURRAN et al. 1991).

As an alternative regulatory mechanism, host cell-specific modifications of viral proteins essential for transcription and replication could be envisaged. Differentially phosphorylated forms of P and V proteins during persistence have been proposed (see above). As shown for vesicular stomatitis virus (VSV), at least one of the functionally required phosphorylations of the P protein is due to the action of a cellular kinase and an inhibition of that kinase activity directly correlates with transcriptional inhibition at least in vitro (BARIK and BANERJEE 1992; SLEAT et al. 1993). Since MV contains two additional phosphorylated proteins, N and V (RIMA 1983; GOMBART et al. 1992), it will be highly interesting to identify the sites of phosphorylation, their usage in different cell types and their functional properties.

3.2.2 Interferon-Dependent Antiviral Activities

As will be pointed out later in this chapter, a characteristic set of cytokines is induced in brain cells infected with MV. Besides the important role of these compounds in intercellular cross-talk, in particular with cells of the immune system, host factors can be induced that may contribute to the regulation of MV gene expression. Since the antiviral actions of type I interferons (IFNS) and IFN inducible proteins have been well documented, these compounds have been discussed to be associated with the establishment of the persistent infection of brain cells. According to one hypothesis, IFN escape variants may spread preferentially in the brain (CARRIGAN and KABAKOFF 1987). However, amongst the IFN inducible proteins, the Mx proteins have been directly related to an antiviral action against influenza virus and VSV, interfering with transcription and/or translation of viral genes (STAEHELI 1990; PAVLOVIC et al. 1990; STAEHELI and PAVLOVIC 1992). Interestingly, expression of Mx proteins has recently been detected in MV-infected brain cell cultures and brain material revealing pronounced restrictions of viral gene expression (KRAUS et al. 1992; SCHNEIDER-SCHAULIES et al. 1993a). A marked down-regulation of MV-specific mRNA synthesis in brain cell cultures constitutively expressing human MxA suggests indeed an important role of that particular protein in contributing to the establishment of a persistent infection (SCHNEIDER-SCHAULIES et al. 1994). The mode of antiviral action of Mx proteins has not been defined yet. Their association with cytoskeletal components (HORISBERGER 1992) and their capacity to bind and hydrolyze GTP, which has been directly linked to their antiviral activity (PITOSSI et al. 1993), led to the proposal that Mx proteins are related to cellular stress proteins (HORISBERGER 1992). In this context it is interesting that cellular stress proteins have been shown to up-regulate canine distemper virus (CDV) gene expression using cell-free assays from mouse neuroblastoma cells (OGLESBEE et al. 1993).

3.2.3 Antibody-Induced Antigenic Modulation

A role of extracellular virus-neutralizing antibodies for regulation of intracellular MV gene expression has been shown in tisue culture experiments using both nonneural and neural cells. In the presence of polyclonal anti-MV antibodies, a selective reduction of MV P and M proteins was observed in infected HeLa cells (FUJINAMI and OLDSTONE 1980). A complete down-regulation of intracellular viral protein expression exclusively following treatment with neutralizing anti-H monoclonal antibodies was observed in persistently MV-infected rat glioma and mouse neuroblastoma cells, but not in persistently infected Vero and lung fibroblast cells (RAGER-ZISMAN et al. 1984; BARRETT et al. 1985; SCHNEIDER-SCHAULIES et al. 1992). Apparently, besides their role in neutralizing free virus particles, antibodies directed against the viral H protein are capable of acting as ligands for their antigen expressed at the surface of neural cells. Subsequently, a so far uncharacterized signal is triggered that ultimately results in a down-regulation of intracellular viral gene expression. In persistently infected rat glioma cells treatment with monoclonal anti-H antibodies led to an immediate, temporary activation of protein kinase C following the induction of a G-protein mediated by inositol triphosphate production (WEINMAN-DORSCH and KOSCHEL 1989). Although the relationship between this activation and the control of MV gene expression has not been established, alterations of cellular kinase activities may well be significant for viral gene expression. In particular, phosphorylation of viral proteins present in the transcription complex might modulate the activity of the latter, as suggested for the P protein of VSV (SLEAT et al. 1993). In fact, the overall efficiency of virus-specific transcription was down-regulated within 24 h following treatment with neutralizing anti-H antibodies in MV persistently infected neural cells. This finding indicates that the activity of the transcription complex may indeed be the target for antibody-induced antigenic modulation (AIAM) in neural cells (SCHNEIDER-SCHAULIES et al. 1992).

Consistent with these findings is the observation that transfer of neutralizing anti-H antibodies to MV-infected newborn rats leads to a subacute CNS disease, whereas the untreated control group succumb to an invariably fatal encephalopathy within a few days (LIEBERT et al. 1990). On the molecular level, expression of the MV envelope proteins was shown to be reduced as a consequence of a significantly restricted expression of viral transcripts. Data obtained by in situ hybridization indicated that the reduced efficiency of viral transcription was due to a restriction at the single cell level rather than reflecting an inhibition of virus spread in the brain. Interestingly, very similar findings were obtained in the presence of high titers of virus-neutralizing antibodies naturally produced in response to the experimental infection in weanling BN rats (LIEBERT et al. 1990; DÖRRIES et al. 1988).

In general, AIAM seems to represent a hightly efficient and important control mechanism for viral gene expression in the brain since observations very similar to the MV system have recently been reported for alphavirus (LEVINE et al. 1991).

3.3. Functional Alterations of Measles Virus-Specific Transcripts

Being transcribed and replicated by an RNA-dependent RNA polymerase, the presence of a considerable amount of sequence variations between MV genomes has to be anticipated, even during lytic infections which are independent of host factors (DOMINGO and HOLLAND 1988). In a lytic infection, natural selection eliminates virus variants loaded with mutations leading to functional impairments. Upon restriction of viral transcription and replication, such as in brain cells, however, compensation of defective gene functions may be inefficient. Single point mutations introduced by the viral polymerase occur apparently independent of the host cell. Also, the editing function seems to be intrinsic to the viral polymerase complex and independent of cellular components, as revealed by the finding that the P mRNA is correctly edited in the absence of cellular proteins in vitro (HORIKAMI and MOYER 1991). Thus, RNA editing is apparently not influenced by the host cell background.

3.3.1 Hypermutation as a Consequnece of a Cellular Duplex RNA-Dependent Adenosine Deaminase

The surprising finding of a hypermutated M gene in a case of MIBE in which 50% of the uridine (U) residues were replaced by cytidine (C) was not attributed to the action of the viral polymerase but rather to that of a cellular enzyme first called

RNA unwindase and now referred to as duplex RNA-dependent adenosine deaminase (DRADA) (CATTANEO et al. 1988; BASS et al. 1989). This assumption was further supported by the fact that all M gene-specific sequences isolated from this particular brain area revealed a high degree of homology, indicating a single step event rather than independently occurring point mutations introduced by the polymerase (CATTANEO et al. 1989).

DRADA has been detected in vitro as an enzyme activity intrinsic to a variety of cell types including brain cells. Within a dsRNA template, adenosine (A) residues are converted to inosine (I) with a frequency up to 50% thereby both destabilizing the dsRNA and altering the informational content of the RNA (WAGNER and NISHIKURA 1988; NISHIKURA et al. 1991). The activity of the dsRNA unwindase, its intracellular location, or both, vary during the cell cycle of mouse fibroblast 3T3 cells (WAGNER and NISHIKURA 1988; WAGNER et al. 1989). In vitro, nuclear extracts of a human neuroblastoma cell line are exceedingly active in modifying a synthetic MV M-specific dsRNA (RATAUL et al. 1992). More recently, the presence of the activity could be demonstrated in cytoplasmic extracts of several neural cell lines (ECKER et al. 1994). Due to its specificity for dsRNA templates and its obvious independence of particular sequence requirements, the activity may contribute substantially to the intracellular antiviral response. In support of this hypothesis, several different hypermutated MV M gene sequences have been encountered in one single SSPE brain (BACZKO et al. 1993).

3.3.2 Translational Control

Besides changing the frequency for the corresponding mRNA, an important site of controlling viral gene expression is undoubtedly translation. Translation may depend on elements intrinsic to the RNA, such as 5' and 3' noncoding sequences and sequence alterations leading to premature termination of the corresponding protein that are independent of the host cell and thus are not considered her. For the M mRNA transcript isolated from the brain of experimentally infected Lewis rats with subacute measles encaphalitis (SAME), translation was restricted in vivo and in vitro, independent of sequence alterations (SCHNEIDER-SCHAULIES et al. 1989).

In contrast, a temperature shift of persistently infected rat glioma cells leads to a selective and reversible translation inhibition of MV M- and F-specific mRNAs, arguing strongly for the involvement of cellular determinants in controlling viral gene functions (OGURA et al. 1987, 1988). Similar observations of a translational inhibition affecting MV protein synthesis either partially or completely have been described as a consequence of in vitro differentiation of tissue culture cells of neural origin (MILLER and CARRIGAN 1982; YOSHIKAWA and YAMANOUCHI 1984; SCHNEIDER-SCHAULIES et al. 1993a). In one of these studies, in vitro differentiation of human glial cell lines prior MV infection led to an almost complete block of viral protein synthesis in vivo, whereas the synthesis of the corresponding mRNAs and their ability to direct the synthesis of translation products in vitro were apparently unaffected (SCHNEIDER-SCHAULIES et al. 1993a). In addition, overall protein synthesis of the infected cells was not inhibited. The obviously specific inhibition of viral rather than cellular gene expression is reminiscent of that described for the antiviral activity of certain IFN-induced proteins (for review see STAEHELI 1990). Although the induction of IFN could be shown by induction of human MxA as a consequence of MV infection in these cell lines, in vitro differentiation did not lead to a detectable stimulation of IFN synthesis, arguing against a pronounced inhibitory effect brought about by phosphatidylinositol (PI) kinase (SCHNEIDER-SCHAULIES et al. 1993a).

4 Alterations of Host Gene Functions of Neural Cells as a Consequence of Measles Virus Infections

Subacute sclerosing panencephalitis begins with subtle mental and behavioral changes and progresses to a massive impact on cellular functions in the brain. In contrast to uninfected brains, numerous MHC class II positive cells can be detected immunohistochemically particularly around blood vessels (HOFMAN et al. 1991). These HLA-DR positive cells have been identified by morphological criteria to be mainly macrophages/microglial cells and reactive astrocytes. Analysis of cytokines revealed the presence of tumor necrosis factor (TNF)-α and IFN-γ and a few interleukin (IL)-1β and IL-2 positive brain cells. The TNF-α positive cells had the morphological appearance of astrocytes, while the IFN-γ positive cells appeared to be macrophages (HOFMAN et al. 1991). These data indicate that microglial cells and astrocytes appear to be activated in SSPE brains and are induced to express MHC class II molecules and certain cytokines. Furthermore, in the CSF of SSPE patients, elevated levels of type I IFNs have been detected (JONCAS et al. 1976). It is not well known how these activated glial cells and their secreted products and the presence of infiltrating lymphocytes influence neuronal cell functions in the brain.

4.1 Effects on Neurotransmitter Signaling

To measure the effect of a MV infection on neural cell functions, in vitro studies were carried out. The level of cAMP, the activity of adenylate cyclase, and the β-adrenergic receptor system were investigated in the persistently MV-infected rat astrocytoma cell line C6/SSPE. It was found that the cAMP response after addition of catecholamines is strongly reduced, that the density of β-adrenergic receptors is decreased by 50%, and that coupling of the receptor to G-proteins is affected (HALBACH and KOSCHEL 1979; KOSCHEL and MÜNZEL 1980). Anti-MV antibodies influence the inositol-phosphate signal transduction pathway in these cells (WEINMANN-DORSCH and KOSCHEL 1989). Furthermore, the Ca^{2+} signal induced by the neurotransmitter endothelin-1 was not detectable in MV persistently infected C6/SSPE cells, since 95% of the binding sites for endothelin-1 were lost (TAS and KOSCHEL 1991). Interestingly, treatment with neutralizing antibodies to MV-H causing antigenic modulation and complete disappearance of viral proteins from these cultures did not lead to reexpression of endothelin-1 binding sites or induction of the Ca^{2+} signal (TAS and KOSCHEL 1991), whereas β-adrenergic receptor-stimulated cAMP synthesis was restored after antibody treatment (BARRETT and KOSCHEL 1983).

Studies with a different family of neurotransmitter receptors, the N-methyl-D-aspartate (NMDA) receptors, revealed an association with MV-induced neuropathology in BALB/c mice. Systemic treatment of mice with the non-competitive NMDA receptor antagonist 5 methyl-10,11-dihydro-5H-dibenzo (a,d) cyclohepten-5, 10, 11-dihydro-5H-dibenzo (a, d) cyclohepten-5, 10-imine maleate (MK801) prevented cellular necrosis and neuronal loss induced by MV infection (ANDERSSON et al. 1990). Thus, MV may have indirect neurodegenerative effects in the brain due to activation of NMDA receptors regulating neuronal survival via Ca^{2+} influx.

4.2 Effects on Cell Surface Molecule Expression

In primary cultures of astrocytes from newborn Lewis rats, MV infection leads to the induction of MHC class II and an increase of MHC class I expression (MASSA et al. 1987). The induction of Ia in these experiments was considerably enhanced after addition of external TNF-α and not mediated by IFN-γ induction, as shown by using antibodies to IFN-γ. The involvement of other cytokines was not investigated in this study and therefore could not be excluded. Recently, the increase of the expression of MHC class I and intercellular adhesion molecule (ICAM-1) on primary rat astrocytes was found to be mediated by type I interferons, the expression of which is induced upon MV-infection as indicated by the presence of Mx proteins in these cultures (KRAUS et al. 1992). However, addition of antibodies to rat type I interferons to virus infected cultures reduced the magnitude of MHC class I and ICAM-1 augmentation only to some extent, suggesting the presence of additional factors. The expression of MHC class I molecules on murine neuroblastoma cells after persistent MV infection was also investigated. In these cells MHC expression was augmented together with the IFN inducible proteins 2–5 oligo A-synthase and dsRNA-activated protein kinase (GOPAS et al. 1992). All these studies indicate that the induction of several cytokines by MV infection may influence both MV replication and host cell functions in a cell type-specific manner.

4.3 Induction of Cytokines in Brain Cells

To investigate the actual set of cytokines induced by MV infection, in vitro studies with different brain cell types were performed. Infection of human astrocytoma cells with MV resulted in a transient expression of a characteristic set of cytokines, namely, IL-1, IFN-β, IL-6 and TNF-α, at the level of mRNA and protein in cell supernatants (SCHNEIDER-SCHAULIES et al. 1993b). Similar cytokines were found after infection of primary rat astrocytes with the neurotropic Newcastle disease virus (NDV; LIEBERMAN et al. 1989) and in rat brain tissue infected by Borna disease virus (BDV) (SHANKAR et al. 1992). In contrast to newly infected cells, MV persistently infected human astrocytoma cells continually produced IL-6 and IFN-β, whereas "NF-α" and IL-1β in most clones were hardly detectable (SCHNEIDER-SCHAULIES et al. 1993b). A similar phenomenon was observed in MV-infected human monocytes (LEOPARDI et al. 1992). However, the pathways for induction of TNF-α and L-1β expression in astrocytoma cells were not suppressed by the MV infection, since TNF-α and IL-1β could still be induced by external stimuli, e.g. diacylglyceroly analog plus calcium ionophore. Under these conditions, persistently infected astrocytoma cells synthesized considerably higher levels of TNF-α and IL-1β than uninfected cells after additional external induction (SCHNEIDER-SCHAULIES et al. 1993b). These results suggest that in MV infections of the CNS a percentage of persistently infected astrocytes may continually synthesize IL-6 and IFN-β, and in the presence of additional external stimuli, as possibly provided by activated lymphocytes, might overexpress the inflammatory cytokines TNF-α and IL-1β . Expression of these cytokines in the brain might decisively influence the disease process of SSPE, since both cytokines have pleiotropic effects in brain tissue (for review see PLATA-SALAMAN 1991).

5 Conclusions and Future Perspectives

The molecular biological studies of MV interaction with brain cells have contributed greatly to elucidate virological aspects of MV infection in SSPE and MIBE. It is now possible to explain why the persistent infection in brain cells is not productive, why membrane changes of infected cells do not occur and why the immune response fails to control the infection or destroy infected brain cells. However, major questions concerning epidemiology and pathogenetic mechanisms are still unanswered. There is no explanation why CNS diseases are so rare in contrast to acute measles, why more boys than girls develop SSPE and why SSPE is more prevalent in rural areas than in large cities. Moreover, the factors determining the long incubation periods of months to years after onset of acute measles and the factors which trigger the disease process are unknown. It would be important to determine how and when measles virus enters the CNS in the course of acute measles. Does measles virus reach the CNS during viremia or by infected lymphocytes or monocytes as observed in CDV infection in dogs (TER MEULEN and STEPHENSON 1983)? Therefore, the characterization of MV infection in lymphocytes and monocytes will be important not only in view of MV-induced immune regulatory changes and lifelong immunity, but also to find out whether latently infected lymphocytes exist in vivo which could, after antigenic stimulation, reach the CNS and carry the virus to brain tissue. Moreover, future studies have to define features underlying MV neurovirulence as suggested by animal studies. In this context, the identification of the MV receptor provides a new and exciting approach to study this aspect. Recently, two prime candidates acting as MV receptors have been identified on nonneural cells, CD46 (NANICHE et al. 1993; DOERIG et al. 1993) and moesin (DUNSTER et al. 1994). CD46 (membrane cofactor protein, MCP) is a member of the "regulators of complement activation" (RCA) protein family that serves critical functions in protecting cells from unspecific complement-mediated lysis by inactivating complement factors (LISZWESKI et al. 1991). Moesin (membrane organizing extension spike protein) belongs to a family of cytoskeletal proteins. It is associated with the inner and outer side of the cell membrane, although it lacks a typical transmembrane domain (FURTHMAYER et al. 1992). So far, the role of these molecules for the process of MV binding and entrance into the target cells is unknown. It will be of interest to characterize the expression of these molecules on subpopulations of brain cells in vivo and to address the consequence of their interactions with MV on brain cell functions. It is obvious that these studies have the potential to provide us with important insights into the pathogenicity of these CNS disorders.

Acknowledgements. We thank Andrea Ecker for generously providing unpublished results and Jens Jorg Schnorr and Armin Schuster for critical reading of the manuscript. Studies cited from our laboratory were supported by the Deutsche Forschungsgemeinschaft and Bundesministerium für Forschung und Technologie.

References

Andersson T, Schultzberg M, Schwarz R, Love A, Wickman C, Kristensson K (1990) NMDA-receptor antagonist prevents measles virus-induced neurodegeneration. Eur J Neurosci 3: 66–71
Baczko K, Lampe J, Liebert UG, Brinckmann U, ter Meulen V, Pardowitz J, Budka H, Cosby SL, Isserte S, Rima BK (1993) Clonal expansion of hypermutated measles virus in a SSPE brain. Virology 197: 188–195
Banerjee AK (1987) Transcription and replication of rhabdoviruses. Microbiol Rev 51: 66–87
Barik S (1992) Transcription of human respiratory syncytial virus genome RNA in vitro: requirement of cellular factor(s). J Virol 66: 6813–6818
Barik S, Banerjee AK (1992) Sequential phosphorylation of the phospho-protein of vesicular stomatitis virus by cellular and viral protein kinases is essential for transcription activation. J Virol 66: 1109–1118
Barrett PN, Koschel K (1983) Effect of antibody-induced modulation of measles (SSPE) virus membrane proteins on β-adrenergic receptor-mediated adenylate cyclase activity. Virology 127: 299–308
Barrett PN, Koschel K, Carter M, ter Meulen V (1985) Effect of measles virus antibodies on a measles SSPE virus persistently infected C6 rat-glioma cell line. J Gen Virol 66: 1411–1421
Bass BL, Weintraub H, Cattaneo R, Billeter MA (1989) Biased hypermutation of viral genomes could be due to the unwinding/modification of the double stranded RNA. Cell 56: 331
Billeter MA, Cattaneo R (1991) Mutations and A/I huypermutations. In: "Measles Virus persistent infections. The paramyxoviruses." Kingsbury D (ed), Plenum, New York, pp. 323–345
Carrigan DR, Kabakoff CM (1987) Identification of a non-productive, cell-associated form of measles virus by its resistance to inhibition by recombinant human interferon. J Virol 61: 1919–1926
Cattaneo R, Rebmann G, Schmid A, Baczko K, ter Meulen V, Billeter MA (1987a) Altered transcripion of a defective measles virus genome derived from a diseased human brain. EMBO J 6: 681–687
Cattaneo R, Rebmann G, Baczko K, ter Meulen V, Billeter MA (1987b) Altered ratios of measles virus transcripts in diseased human brains. Virology. 160: 523–526
Cattaneo R, Schmid A, Eschle D, Baczko K, ter Meulen V, Billeter MA (1988) Biased hypermutation and other genetic changes in defective measles viruses in human brain infections. Cell 55: 255–265
Cattaneo R, Schmid A, Spielhofer P, Kaelin K, Baczko K, ter Meulen V, Pardowitz J, Flanagan S, Rima BK, Udem SA, Billeter MA (1989) Mutated and hypermutated genes of persistent measles viruses which caused lethal human brain diseases. Virology 173: 415–425

Curran J, Boeck R, Kolakofsky D (1991) The Sendai virus P gene expresses both an essential protein and an inhibitor of RNA synthesis by shuffling modules via mRNA editing. EMBO J 10: 3079–3085
De BP, Burdsall AL, Banerjee AK (1993) Role of cellular actin in human parainfluenza virus type 3 genome transcription. J Biol Chem 268: 5703–5710
Döerig RE, Marcil A, Chopra A, Richardson CD (1993) The human CD46 molecule is a receptor for measles virus (Edmonston strain). Cell 75: 295–305
Dörries R, Liebert UG, ter Meulen V (1988) Comparative analysis of virus-specific antibodies and immunoglobulins in serum and cerebrospinal fluid of subacute measles virus induced encephalomyelitis (SAM) in rats and subacute sclerosing panencephalitis (SSPE). J Neuroimmunol 19: 339–352
Domingo E, Holland JJ (1988) High error rates, population equilibrium and evolution in RNA replication systems. In: genetics. Domingo E. Holland J.J. Ahlquist P. (eds) RNA genetics. CRC Press, Boca Raton, pp 3–36
Dunster LM, Schneider-SchauliesJ, Loffler S, Lankes W, Schwarz- Albiez R, Lottspeich F, ter Meulen V (1994) Moesin: a cell membrane protein linked with suceptibility to measles virus infection. Virology 198: 265–274
Ecker A, ter Meulen V, Baczlco V, and Schneider-Schaulies S (1994) Measles virus-specific dsRNAS are targets for unwinding/modifying activity in neural cells in vitro. Submitted.
Fujinami RS, Oldstone MBA (1980) Altertions in expression of measles virus polypeptides by antibody: molecular events in antibody induced anigenic modulation. J Immunol 125: 78–85
Furthmayer HW, Lankes W, Amieva A (1992) Moesin, a new cytoskeletal protein and constituent of filopodia: its role in cellular functions. Kidney Int 41: 665–670
Gombart AF, Hirano A, Wong TC (1992) Expression and properties of the V protein in acute measles virus and subacute sclerosing panencephalitis viurs strains. Virus Res 25: 63–78
Gopas J, Yitzhaki D, Segev Y, Salzberg S, Trink B, Isakov N, Reager-Zisman B (1992) Persistent measles virus infection enhances MHC class I expression and immunogenicity of murine neuroblastoma cells. Cancer Immunol Immunother 34: 313–320
Halbach M, Koschel K (1979) Impairment of hormone dependent signal transfer by chronic SSPE virus infection. J Gen Virol 42: 615–619
Hofman FM, Hinton DR, Baemayr J, Weil M, Merrill JE (1991) Lymphokines and immunoregulatory molecules in subacute sclerosing panencephalitis. Clin Immunol Immunopathol 58: 331–342
Horikami SM, Moyer SA (1991) Synthesis of leader RNA and editing of the P mRNA during transcription by purified measles virus. J Virol 65: 5342–5347
Horikami SM, Curran J, Kolakofsky D, Moyer SA (1992) Complexes of the Sendai virus NP/P and P/L proteins are required for defective interfering particle genome replication in vitro. J Virol 66: 4901–4908
Horisberger MA (1992) Interferon-induced human protein MxA is a GTPase which binds transiently to cellular proteins. J Virol 66: 4705–4709
Joncas JH, Robillard LR, Boudreault A, Leyritz M, McLaughlin, BJM (1976) Interferon in serum and cerebrospinal fluid in subacute sclerosing panencephalitis. Can Med Assoc J 115: 309
Koschel K, Munzel P (1980) Persistent paramyxovirus infections and behavior of β-adrenergic receptors in C6 rat glioma cells. J Gen Virol 47: 513–517
Kraus E, Schneider-Schaulies S, Miyasaka M, Tamatani T, Sedgwick J (1992) Augmentation of major histokompatibility complex class I and ICAM-1 expression on glial cells following measles virus infection: evidence for the role of type-1 interferon. Eur J Immunol 22: 175–182
Leopardi R, Vanionpää R, Hurme M, Siljander P, Salmi A (1992) Measles virus infection enhances IL-1b but reduced tumor necrosis factor-α expression in human monocytes. J Immunol 149: 2397–2401
Leopardi R, Hukkanen V, Vanionpaa R, Salmi AA (1993) Cell proteins bind to sites within the 3 noncoding region and the positive strand leader sequence of measles virus RNA. J Virol 67: 785–790
Levine B, Hardwick JM, Trapp BD, Crawford TO, Bollinger RC, Griffin DE (1991) Antibody mediated clearance of alphavirus infection from neurons. Science 254: 856–860
Lieberman AP, Pitha PM, Shin HS, Shin ML (1989) Production of tumor necrosis factor and other cytokines by astrocytes stimulated with lipopolysaccharide or a neurotropic virus. Proc Natl Acad Sci USA 86: 6348–6352
Liebert UG, ter Meulen V (1987) Virological aspects of measles virus induced encephalomyelitis in Lewis and BN rats. J Gen Virol 68: 1715–1722
Liebert UG, Schneider-Schaulies S, Baczko K, ter Meulen V (1990) Antibody-induced restriction of viral gene expression in measles encephalitis in rats. J Virol 64: 706–713
Liszweski MK, Post TW, Atkinson JP (1991) Membrane cofactor protein (MCP or CD46): Newest member of the regulators of complement activation gene cluster. Annu Rev Immunol 9: 431–455

Massa PT, Schimpl A, Wecker E, ter Meulen V (1987) Tumor necrosis factor amplifies measles virus mediated la induction on astrocytes. Proc Natl Acad Sci USA 84: 7242–7245
Miller CA, Carrigan DR (1982) Reversible repression and activation of measles virus infection in neural cells. Proc Natl Acad Sci USA 79: 1629–1633
Moyer SA, Baker Sc, Lessard JL (1986) Tubulin: a factor necessary for the synthesis of both Sendai virus and vesicular stomatitis virus RNA. Proc Natl Acad Sci USA 83: 5405–5409
Moyer Sa, Baker SC, Horikami SM (1990) Host cell proteins required for measles virus reproduction. J Gen Virol 71: 775–783
Naniche D, Varior-Krishnan G, Cervoni F, Wild TF, Rossi B, Rabourdin-Combe C, Gerlier D (1993) Human membrane cofactor protein (CD46) acts a cellular receptor for measles virus. J Virol 67: 6025–6032
Nishikura K, Yoo C, Kim U, Murray JM, Estes PA, Cash FE, Liebhaber SA (1991) Substrate specificity of the dsRNA unwinding/modifying activity. EMBO J: 3523–3432
Oglesbee MJ, Kenney H, Kenny T, Krakowa S (1993) Enhanced production of morbillivirus gene specific RNAs following induction of the cellular stress response in stable persistent infection. Virology 192: 556–567
Ogura H, Baczko K, Rima BK, ter Meulen V (1987) Selective inhibition of translation of the mRNA coding for the measles virus membrane protein at elevated temperatures. J Virol 61: 472–479
Ogura H, Rima BK, Baczko K, ter Meulen V (1988) Restricted synthesis of the fusion protein of Measles Virus at elevated temperatures. J Gen Virol 69: 925–929
Pavlovic J, Zurcher T, Haller O, Staeheli P (1990) Resistance to influenza virus and vesicular stomatitis virus conferred by expression of human MxA protein J Virol 64: 3370–3375
Pitossi F, Blank A Schroder A, Schwarz A, Hussi P, Schwemmle M, Pavlovic J, Staeheli P (1993) A functional GTP binding motif is necessary for antiviral activity of mx proteins. J Virol 67: 6726–6732
Plata-Salaman CR (1991) Imunoregulators in the central nervous system. Neurosci Biobehav Rev 15: 185–215
Rager-Zisman B, Egan JE, Kress Y, Bloom BR (1984) Isolation of Cold-Sensitive mutatnts of Measles Virus from persistently infected murine neuroblastoma cells. J Virol 51: 845–855
Rammohan KW, McFarland MF, McFarlin DE (1981) Induction of subacute murine measles encephalitis by monoclonal antibody to virus hemagglutinin. Nature 290: 588–589
Rataul SM, Hirano A, Wong TC (1992) Irreversible modification of measles virus RNA in vitro by nuclear RNA-unwinding activity in human neuroblastoma cells. J Virol 66: 1769–1773
Rima BK (1983) The proteins of morbilliviruses. J Gen Virol 64: 1205–1219
Schneider-Schaulies S, ter Meulen V (1992) Molecular Aspects of Measles Virus induced central nervous system diseases. In: Roos RP (ed) Molecular neurovirology. Humana, Clifton, pp. 419–449
Schneider-Schaulies S, Liebert UG, Baczko K, Cattaneo R, Billeter M, ter Meulen V (1989) Restriction of measles virus gene expression in acute and subacute encephalitis of Lewis rats. Virology 171: 525–534
Schneider-Schaulies S, Liebert UG, Baczko K, ter Meulen V (1990) Restricted expression of measles virus in primary rat astroglial cells. Virology 177: 802–806
Schneider-Schaulies S, Liebert UG, Segev Y, Rager-Zisman B, Wolfson M, ter Meulen V (1992) Antibody-dependent transcriptional regulation of measles virus in persistently infected neural cells. J Virol 66: 5534–5541
Schneider-Schaulies S, Schneider-Schaulies J, Bayer M, Loffler S, ter Meulen V (1993a) Spontaneous and differentiation dependent regulation of measles virus gene expression in human glial cells. J Virol 67: 3375–3383
Schneider-Schaulies J, Schneider-Schaulies S, ter Meulen V (1993b) Differential induction of cytokines by primary of presistent measles virus infections in human glial cells. Virology 195: 219–228
Schneider-Schaulies S, Schneider-Schaulies J, Schuster A, Bayer M, Pavlovic J, ter Meulen V (1994) Cell type specific MAX-mediated inhibition of Measles virus Transcription in human brain cells. J Virol, in Press
Shankar V, Kao M, Hamir AN, Sheng H, Koprowski H, Dietszschold B (1992) Kinetics of virus spread and changes in levels of several cytokine mRNAs in the brain after intranasal infection of rats with Borna disease virus. J Virol 66: 992–998
Sleat DE, Chikkala NF, Gautam S, Banerjee AK (1993) Restricted replication of vesicular stomatitis virus in T lymphocytes is coincident with a deficiency in a cellular protein kinase required for viral transcription. J Gen Virol 73: 3125–3132
Staeheli P (1990) Interferon. induced proteins and the antiviral state. Adv Virus Res 38: 147–200
Staeheli P, Pavlovic J (1992) Inhibition of Vesicular stomatitis virus mRNA synthesis by human MxA protein. J Virol 65: 4498–4501
Swoveland PT, Johnson KP (1989) Host age and cell type influence measles virus protein expression in the central nervous system. Virology 170: 131–138

Tas P, Koschel K (1991) Loss of the endothelin signal pathway in C6 rat glioma cells persistently infected with measles virus. Proc Natl Acad Sci USA 88: 6736–6739

Ter Meulen V, Stephenson JR (1983) The posible role of viral infections in MS and other related demyelinating diseases. In: . Hallpike JF, Adams CWM, Tourtellotte WW (eds) Multiple sclerosis Champany and Hall, London pp 241–274

Wagner RW, Nishikura K (1988) Cell cycle expression of RNA duplex unwindase activity in mammalian cells. Mol Cell Biol 8: 770–777

Wagner RW, Smith JE, Cooperman BS, Nishikura K (1989) A double stranded RNA unwinding activity introduces structural alterations by means of adenosine to inosine conversions in mammalian cells and Xenopus eggs. Proc Natl Acad Sci USA 86: 2647–2651

Weinmann-Dorsch C, Koschel K (1989) Coupling of viral membrane proteins to phosphatidylinositide signalling sytem. FEBS Lett 247: 185–188

Yoshikawa Y, Yamanouchi K (1984) Effects of papaverine treatment on replication of measles virus in human neural and non-neural cells. J Virol 50: 489–495

Immune Responses During Measles Virus Infection

D.E. GRIFFIN

1 Introduction

The immune responses elicited by measles virus (MV) in many ways define the epidemiology and the illnesses produced by infection. During a study of the 1886 epidemic of measles in the Faroe Islands, the young Danish physician Peter Panum identified not only the highly infectious nature of the virus, but also the 14 day incubation period before the onset of the rash, the severity of the disease in infants and the presence of lifelong immunity in the absence of reexposure in elderly residents (PANUM 1938). In 1908 another observant physician, von Pirquet,

---

Departments of Medicine and Neurology, Meyer 6-181, Johns Hopkins University School of Medicine, Baltimore, MD 21287, USA

reported the loss of skin test reactivity and the reactivation of tuberculosis after measles that occurred in the residents of a Viennese tuberculosis sanitarium (VON PIRQUET 1908). In 1790 James Lucas, a surgeon from Leeds, described the first case of post-measles encephalomyelitis (LUCAS 1790). Thus, by early in the twentieth century most of the features of the disease that we now seek to understand, using the knowledge and tools of modern immunology, had been established.

2 Overview of the Immune Response to Natural Measles

Measles virus is spread from one person to another by the respiratory route. During the incubation period the virus spreads from the respiratory mucosa and lungs to local lymphoid tissue, where virus replication occurs primarily in macrophages (ESOLEN et al. 1993; MOENCH et al. 1988). A viremia is established that allows systemic spread of the virus to other lymphoid tissues and to endothelial and epithelial cells in multiple organs including the conjunctivae, skin and liver. It is probable that the prodromal symptoms of fever, cough and conjunctivitis that occur approximately 14 days after infection are the earliest indicators of the induction of MV-specific immunity. The rash marks the clear onset of the immune response to infection as determined by the appearance of antiviral antibody. The immune response is also marked by infiltration of mononuclear cells into local areas of virus replication and the appearance of virus-specific T cells in the blood (Fig. 1). Virus is rapidly cleared from blood and tissue after the appearance of the rash. Individuals with severe defects in cellular immunity may develop progressive MV infection in the absence of a rash) (ENDERS et al. 1959). Paradoxically, the development of this immune response, which is highly effective in clearing virus and in establishing long-term immunity, is associated with numerous abnormalities of immune function, which contribute to the morbidity and mortality of measles. Delayed type hypersensitivity (DTH) skin test responses to recall antigens disappear (VON PIRQUET 1908; TAMASHIRO et al. 1987), there is an increased

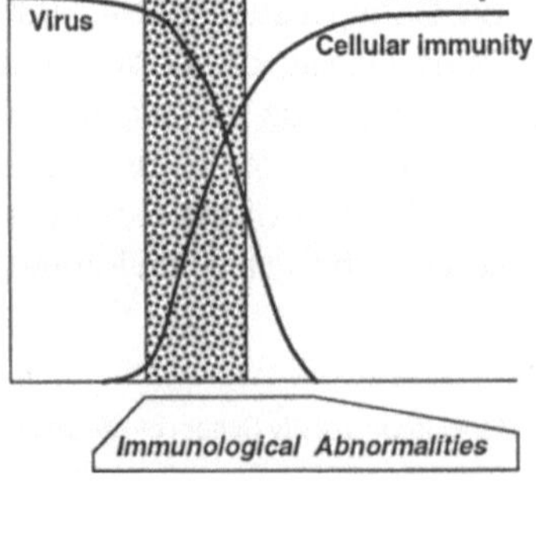

Fig. 1. Overview of the time course of the immune responses during natural MV infection

susceptibility to secondary infections (BECKFORD et al. 1985; MORELY 1969) and autoimmune encephalomyelitis may occur (MILLER 1964; LITVAK et al. 1943). The lack of a practical animal model that mimics human disease means that knowledge of the immunologic contributions to measles and its complications have come almost exclusively from the human studies that will be described in this review.

3 Antibody Responses

3.1 Time course and Isotypes

Measles Virus antibody is first detectable at the onset of the rash and titers rise rapidly thereafter (BECH 1959). Initial immunoglobulin isotypes produced are IgM and IgA followed later by IgG (Schluederberg 1965). The IgG response is primarily of the IgG1 and IgG4 isotypes (MATHIESEN et al. 1990; EHRNST 1978). IgG1 is efficiently transported across the placenta so that titers of measles antibody in cord blood are often higher than in the maternal circulation (BLACK 1989). Plasma IgE levels also rise, but there is no indication that the IgE contains MV-specific antibody (GRIFFIN et al. 1985).

3.2 Specificity

Antibody is induced to most viral proteins (NORRBY et al. 1981; STEPHENSON and TER MEULEN 1979; GRAVES et al. 1984 (Fig. 2). Antiviral antibodies can be measured by their biological effects (e.g., neutralization, complement fixation, inhibition of hemagglutination, etc.) or by their reactions with proteins in immunoprecipation, enzyme immunoassays or western blot assays. The most abundant and rapidly produced antibody is to the nucleocapsid (N) protein, and this is the major antibody detected in the complement fixation test (NORRBY and GOLLMAR 1972). The COOH-terminal portion of the N protein is variable (TAYLOR et al. 1991) and two antigenic sites (residues 457-476 and 518-525) have been mapped to this region using monoclonal antibodies. A conserved epitope (residues 122-150) is present in the NH_2-terminal region of the protein (BUCKLAND et al. 1989).

Antibody to the hemagglutinin (H) protein is measured biologically by inhibition of MV-induced agglutination of monkey erythrocytes (HI test) (NORRBY and GOLLMAR 1972). Antibodies to the H protein are also the primary antibodies measured by serum neutralization of virus infectivity in tissue culture. In general, neutralization and HI titers correlate well (BLACK 1989). Antigenic differences in H exist between MV strains (TRUDGETT et al. 1981; TER MEULEN et al. 1981). Four overlapping epitopes have been identified by monoclonal antibody reactivity but the specific location of these antigenic regions has not yet been mapped on the H protein (SHESHBERADARAN and NORRBY 1986; CARTER et al. 1982). Neutralizing

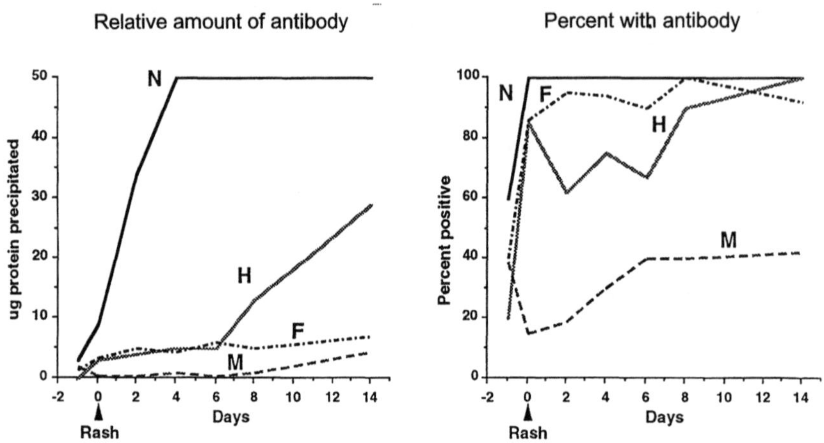

Fig. 2. The time courses of the appearance of antibody to individual MV proteins H, hemagglutinin; N, nucleocapsid; F, fusion; M, matrix) at various times after appearance of the rash (Adapted, with permission, from GRAVES et al. 1984)

antibody is thought to play a prominent role in preventing reinfection. Therefore, HI or neutralization tests are most often used to evaluate vaccine responses and to assess susceptibility to measles.

Antibody to the fusion (F) protein is measured biologically by inhibition of MV-induced hemolysis of monkey erythrocytes (HLI). The F protein is well conserved between MV strains (ROTA et al. 1992). Monoclonal antibodies define at least two conformational epitopes (MALVOISIN and WILD 1990). Mapping of linear epitopes using peptides has shown that epitopes on F are diverse and scattered along the protein (WIESMULLER et al. 1992). Antibody to the F protein probably contributes to virus neutralization by preventing fusion of the virus membrane with the cell membrane (MALVOISIN and WILD 1990) and may therefore limit spread of infection to new cells (MERZ et al. 1980). Antibodies to both F and H can direct complement-mediated lysis of infected cells (EHRNST 1975).

The M protein appears to be less immunogenic than N, H and F with only a portion of individuals making an antibody response, even during chronic infection (GRAVES et al. 1984; NORRBY et al. 1981; MACHAMER et al. 1980) (Fig 2). The potential contribution of this antibody to measles pathogenesis is unknown although increased antibody to M is associated with atypical measles (MACHAMER et al. 1980; NORRBY et al. 1981).

3.3 Role in Infection

The best evidence for the importance of antibody in preventing infection is the protection from infection of infants with maternal antibody (BLACK 1989; HALSEY et al. 1985; ALBRECHT et al. 1977). In general, there is a strong correlation between levels of neutralizing antibody and protection (SIBER et al. 1993; ALBRECHT et al. 1977).

Antibody is also capable of preventing virus-induced membrane fusion (Wild et al. 1991), of lysing infected cells (Ehrnst 1975), of modulating viral antigens from the cell surface (Barrett et al. 1985; Joseph and Oldstone 1975) and of suppressing intracellular synthesis of virus protein and RNA in vitro (Fujinami and Oldstone 1979; Schneider-Schaulies et al. 1992). Therefore, antibody may also play a role in controlling viral replication and also in establishing persistent infection (Rammohan et al. 1983).

3.4 Longevity

The reason for the longevity of the antibody response to MV is not known. Although reexposure may boost this response it is not required (Panum 1938). There is no evidence to date for persistence of latent virus, however, it is possible that viral antigens may persist in the absence of continued virus replication. Follicular dendritic cells trap immune complexes and are able to retain antigen in lymphoid tissue for prolonged periods. These highly convoluted, antigen-retaining cells are found in germinal centers of spleen and lymph nodes where they stimulate memory B cells and appear essential for long-term maintenance of antibody responses (Donaldson et al. 1986; Tew et al. 1990; Gray and Skarvall 1988). The extensive replication of MV in lymphoid tissue may maximize the interaction of viral antigen with follicular dendritic cells leading to retention of viral antigen in multiple lymphoid organs and long-term stimulation of MV-specific IgG (Donaldson et al. 1986).

4 Cellular Immune Responses

4.1 Nonspecific

Both nonspecific and specific cellular immune responses may be activated during systemic virus infections. Although induction of the synthesis of α- and β-interferon (IFN) and subsequent activation of natural killer cells are early responses to many viral infections there is little evidence that these are important components of the nonspecific response to measles. IFN levels in plasma are not elevated (Shiozawa et al. 1988) and natural killer cell activity is lower than it is in control individuals without infection (Griffin et al. 1990b).

4.2 Specific

There is abundant evidence for activation of T cell-driven cellular immune responses during measles. Plasma levels of the soluble forms of the T cell surface molecules CD4, CD8, IL-2 receptor and β_2 microglobulin are elevated (Griffin and Ward 1993; Griffin et al. 1989, 1992) as are the T cell products interleukin (IL)-2,

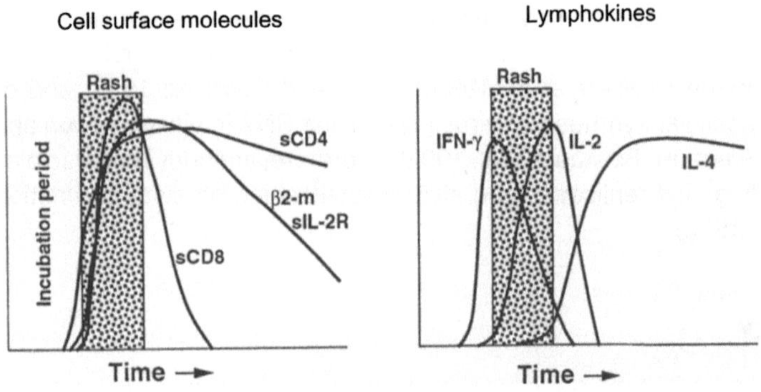

Fig. 3. The levels of soluble (s) cell surface molecules and lymphokines in the plasma at various times during measles, *β2m*, β2 microglobulin, *IL-2R*, interleukin-2 receptor; *IFN*, interferon

IL-4 and IFN-γ (GRIFFIN et al. 1990b; GRIFFIN and WARD 1993). The time courses of these plasma elevations are distinctive and provide some insights into the relative contributions of different components of the cellular immune response at various times during disease and recovery from disease (Fig. 3).

Evidence for activation of T cells becomes apparent during the prodromal phase of the illness when plasma levels of IFN-γ and soluble IL-2 receptor are increased (GRIFFIN et al 1989, 1990b). Therefore, it is likely that the prodromal symptoms of measles are associated with the earliest phases of the immune response to infection, but full expression of this response occurs with the onset of the rash. Fever peaks during the first 24–48 hours of the rash, antibody becomes detectable and mononuclear cells infiltrate sites of virus replication, including the skin.

4.3 CD8

Measles virus specific $CD8^+$ cytotoxic T lymphocytes (CTLs) (VAN BINNENDIJK et al. 1990; KRETH et al. 1979) and proliferating CD8 T cells (WARD et al. 1990) are present in the blood. Soluble CD8, a byproduct of the interaction of activated $CD8^+$ T cells with target cells, is increased in plasma (GRIFFIN et al. 1989) (Fig. 3). CD8 T cells recognize viral antigens after the virus proteins have been synthesized, processed to peptides and presented on the surface of infected cells in the context of (complexed with) major histocompatibility complex (MHC) class I molecules. β_2 microglobulin, the class I molecule constant chain, is also increased in plasma for a prolonged period (GRIFFIN et al. 1992) (Fig 3). Lymphocytes and monocytes are found within the rash. Although the types of cells have not been identified, it is presumed that CD8 T cells are an important component of the lymphocytic infiltrates found at the sites of virus replication and eliminate infected cells by class I-restricted cytotoxic mechanisms. IFN-γ, an important product of cytotoxic CD8 T cells, is also elevated in plasma during this time (GRIFFIN et al.

1990a) (Fig. 3). Long after recovery from measles memory CD8 CTLs can be demon-strated in peripheral blood mononuclear cells (PBMCs) (LUCAS et al. 1982; VAN BINNENDIJK et al. 1989).

There is some suggestion that severity of disease may be linked to certain class I haplotypes (COOVADIA et al. 1981). A HLA B27-restricted CD8 T cell epitope has been identified on the F protein, near the transmembrane segment (residues 438–446) of the protein (VAN BINNENDIJK et al. 1993). Class I-restricted epitopes are no doubt present on other MV proteins but have not yet been identified.

4.4 CD4 and Cytokines

CD4 T cells are also activated in response to MV infection. CD4 T cells are proliferating during the rash (WARD et al. 1990) and soluble CD4 becomes elevated and remains elevated for several weeks (GRIFFIN and WARD 1993) (Fig. 3). However, the two parameters most often used to assess CD4-dependent T cell immunity, DTH skin test responses and in vitro proliferation to viral antigens, are often not demonstrable or are only minimally reactive in measles (GREENSTEIN and MCFARLAND 1983; FULGINITI and ARTHUR 1969). MV-specific MHC class II-restricted cytolytic CD4 cells are generated during measles (VAN BINNENDIJK et al. 1990; JACOBSON et al. 1984) and may contribute to virus clearance from tissue. However CD4 T cells probably exert most of their influence through local secretion of cytokines resulting in attraction, activation and deactivation of macrophages and in proliferation and differentiation of B cells and T cells.

Functional types of CD4 T cells can be distinguished by the types of lymphokines they produce. After initial stimulation by antigen, CD4 cells (Th0) produce abundant IL-2 and are capable of producing many other lymphokines. After restimulation, memory CD4 T cells that produce IL-2, IFN-γ and tumor necrosis factor (TNF)-β (type 1) can be distinguished from those producing IL-4, IL-5 and IL-10 (type 2). These definitions are based on study of T cell clones and, for many T cells, these profiles are overlapping. Nevertheless, CD4 T cell function is largely determined by the cytokines the cells produce. Thus, type 1 CD4 T cells (Th1) are important for macrophage activation leading to DTH responses (IFN-γ), lymphocyte proliferation (IL-2) and MHC class II-restricted cytotoxicity (TNF-β) while type 2 cells (Th2) are important for macrophage deactivation (IL-4 and IL-10) and B cell help (IL-4, IL-5, IL-10) (MOSMANN and COFFMAN 1987).

Measurement of cytokines released into the plasma during measles shows an increase in IFN-γ (a product of CD8 cytotoxic cells and Th1 CD4 cells) during the rash (GRIFFIN et al. 1990a) (Fig. 3). As the rash subsides plasma levels of IL-4 (product of Th2 cells) rise and remain elevated for several weeks (Fig. 3) (GRIFFIN and WARD 1993). This pattern of cytokine production suggests activation of CD8 cells (soluble CD8 and IFN-γ) and of Th0 CD4 cells (soluble CD4 and IL-2) during the rash followed by a more prolonged activation of CD4 cells producing type 2 cytokines (soluble CD4 and IL-4) after resolution of the rash.

Studies of cultured PBMCs from patients with measles have shown spontaneous production of soluble IL-2 receptor indicating in vivo activation of T cells. Stimulation of PBMCs with T cell mitogens such as phytohemagglutinin or antibody to CD3 elicits little IFN-γ or IL-2, but large amounts of IL-4 (WARD et al. 1991; GRIFFIN and WARD 1993).

The lack of long-term activation of cells producing type 1 cytokines (IL-2 and IFN-γ) may in part explain the observation that MV-specific DTH skin test responses do not develop (NELSON et al. 1966; FULGINITI and ARTHUR 1969) and that MV-specific lymphoproliferative responses are low (GREENSTEIN and MCFARLAND 1983; HIRSCH et al. 1984). Long-term activation of cells producing IL-4 may contribute to the polyclonal B cell activation and increased IgE production during the acute phases of measles (GRIFFIN et al. 1985) and to the predominance of MV-specific IgG1 and IgG4 antibody (MATHIESEN et al. 1990).

The MV proteins that can stimulate proliferation of MHC class II-restricted CD4 T cells include H, N, P, F and M (ROSE et al. 1984; VAN BINNENDIJK et al. 1992; BELLINI et al. 1981; ILONEN et al. 1990; JACOBSON et al. 1989). Cytotoxic CD4 T cells have been reported that recognize F (VAN BINNENDIJK et al. 1993). Specific epitopes have been mapped on F where three different class II-restricted CTL clones recognize sequences clustered between amino acids 379 and 466 (VAN BINNENDIJK et al. 1993). Class II-restricted epitopes on other proteins have not yet been identified.

5 Immunologic Abnormalities

5.1 Immune Suppression

The immune suppression that accompanies measles is thought to contribute to the susceptibility to secondary infections that accounts for most of the morbidity and mortality associated with measles (MILLER 1964; MORLEY 1969; BECKFORD et al. 1985). There is both in vivo and in vitro evidence of this immune suppression. In vivo, DTH skin test responses to recall antigens, such as tuberculin, are suppressed for several weeks after apparent recovery from infection (VON PIRQUET 1908; TAMASHIRO et al. 1987). In vitro, lymphoproliferative responses to mitogens are suppressed (HIRSCH et al. 1984; ARNEBORN and BIBERFELD 1983) and lymphokine production is abnormal (CRESPI et al. 1988; WARD et al. 1991; JOFFE and RABSON 1981). This immune suppression is occurring in the context of substantial immune activation (Fig. 3) (ARNEBORN and BIBERFELD 1983; WARD et al. 1990; GRIFFIN et al. 1989; 1990a, 1992; GRIFFIN and WARD 1993; AICARDI et al. 1977; FURUKAWA et al. 1992) and the induction of a MV-specific immune response that is effective both in clearance of virus and in induction of long-term immunity. Analysis of the in vitro abnormalities has suggested defects in the responses of both monocytes (the antigen-presenting cells and producers of accessory molecules such as IL-1 and TNF-α) and lymphocytes (the cells actually proliferating and responsible for production of IL-2).

Monocytes are infected during measles (Esolen et al. 1993). Infection may lead to cell lysis and/or functional alteration. IL-1 production in vitro is generally normal to elevated in PBMCs from measles patients (Ward et al. 1991) and the addition of supplemental IL-1 to lymphoproliferation assays rarely helpful (Griffin et al. 1987; Ward et al. 1991). However, in vitro TNF-α production by PBMCs is low boh during acute and convalescent phases of the disease (Ward et al. 1991) and plasma levels of TNF-α never rise (Furukawa et al. 1992). Deficiency of this proinflammatory cytokine may contribute to decreased lymphoproliferation in vitro and may blunt responses to new infectious challenges in vivo. Both increased IL-1 and decreased TNF-α are observed when macrophage lineage cells are infected with MV in vitro (Leopardi et al. 1992) suggesting that these abnormalities result directly from infection. These in vitro-infected macrophages have increased expression of MHC class II and can present MV antigens efficiently but are defective in presentation of other antigens (Leopardi et al. 1993). The effect of age of infection on duration and severity of monocyte dysfunction in vivo is not known, but monocytes from cord blood are more susceptible to MV infection than monocytes from adults (Sullivan et al. 1975).

T cell numbers are decreased during measles but normal proportions of CD4 and CD8 cells are maintained (Arneborn and Biberfeld 1983; Ward et al. 1990). T cells can be infected with MV in vitro if stimulated to proliferate (Sullivan et al. 1975; Joseph and Oldstone 1975; Salonen et al. 1988) but T cells have not been directly demonstrated to be infected in vivo. Therefore, abnormalities in T cell responses may be secondary to other changes occurring during infection. One of the reasons for low in vitro T cell proliferation to mitogens, especially during convalescence, is inadequate production of IL-2.

Addition of recombinant IL-2 to cultures supports spontaneous proliferation and improves mitogen responses (Ward and Griffin 1993; Griffin et al. 1986). When in vitro production of other cytokines is assessed, IFN-γ is low to normal and IL-4 is high compared to controls (Griffin and Ward 1993; Crespi et al. 1988; Ward et al. 1991) suggesting that type 2 T cells are preferentially activated. Neutralization of IL-4 in vitro improves proliferation (Griffin and Ward 1993). IL-4 serves as a general suppressant of macrophage activation and DTH responses (Martinez et al. 1990; Larner et al. 1993) and may be one of the mechanisms by which in vivo tuberculin responses and in vitro lymphoproliferative responses are suppressed during measles. These abnormalities of macrophage and T cell function may increase susceptibility to other infections.

5.2 Autoimmune Encephalomyelitis

Post-measles encephalomyelitis complicates 1 in 1000 cases of measles, mostly in older individuals (Miller 1964) and usually occurs within 2 weeks after the onset of the rash (Litvak et al. 1943). Pathologic studies show that this is a perivenular demyelinating disease rather than an inclusion body encephalitis (Litvak et al. 1943). There is little evidence for virus in the brain either as assessed directly by virus isolation, antigen detection or viral RNA detection or indirectly by appearance of MV-specific antibody in the cerebrospinal fluid (RUCKLE and ROGERS 1957; JOHNSON et al. 1984; GENDELMAN et al. 1984). There is, however, an induction of an immune response to myelin basic protein suggesting an autoimmune disease (JOHNSON et al. 1984) similar to experimental autoimmune encephalomyelitis. Patients with encephalomyelitis differ from uncomplicated patients by having more marked and prolonged immunologic abnormalities than patients with uncomplicated disease. In particular, IgE is more persistently elevated and soluble IL-2 receptor is lower in those patients with encephalitis (GRIFFIN et al. 1985, 1989). The timing of this complication suggests that immune activation and dysregulation due to MV infection may play a role in the induction of this disease. A similar disease has been reported as a complication of IL-2 infusion (GORETSKY 1990).

6 Vaccine-Induced Immunity

6.1 Live Attenuated Vaccine

The measles vaccine in current use around the world is an attenuated live virus vaccine administered subcutaneously usually between the ages of 9 and 20 months. Antibody responses, as measured by HI, virus neutralization and ELISA, are good and protect against measles as long as the vaccine is given after maternal antibody has declined (ALBRECHT et al. 1977). As with natural infection, the isotype of MV-specific antibodies in plasma is primarily IgG1 (MATHIESEN et al. 1990). The level of antibody induced is generally lower than after natural infection and may decay somewhat more rapidly but is still measurable in most individuals 10-15 years after immunization in the absence of boosting infections (BLACK 1989; KING et al. 1993; MARKOWITZ et al. 1990).

Cellular immune responses to vaccine have received limited study, but are assumed to be elicited in a manner similar to natural infection since antigen presenting cells will be infected and able to present processed MV proteins in the context of MHC class I (generally requires intracellular protein synthesis) as well as class II and therefore should stimulate both MV-specific CD8 and CD4 T cells. As with natural infection, MV-specific DTH responses are not stimulated and proliferative responses are less than robust (FULGINITI and ARTHUR 1969). CTL responses elicited after restimulation with MV-infected cells in vitro are considerably lower than after natural infection (WU et al. 1993). This observation is of potential importance for secondary vaccine failure.

Recently, an increased mortality extending over several years has been reported in infants receiving a high-titered ($\geq 10^{4.7}$ pfu/dose) measles vaccine between 4 and 8 months of age (GARENNE et al. 1991). Since deaths were due to a variety of causes, primarily other infections, the capacity of live attenuated measles vaccine to induce immunologic abnormalities analogous to natural infection has been an area of interest. Studies of older children and adults have shown that administration of measles vaccine is associated with decreased numbers of circulating lymphocytes (MANDALENAKI-ASFI et al. 1976), loss of DTH skin test responses to recall antigens, decreases in vitro proliferative responses to mitogens (HIRSCH et al. 1981; FIREMAN et al. 1969; ZWEIMAN et al. 1971) and abnormalities of cytokine production, including increased synthesis of IL-4 (HIRSCH et al. 1981; WARD and GRIFFIN 1993). It is possible, but has not been demonstrated, that these abnormalities are more pronounced or prolonged with a higher dose of vaccine virus.

6.2 Inactivated Vaccine

Simultaneously with the development of the live virus vaccine in the early 1960s Tween/ether-inactivated and formalin-inactivated, alum-precipitated measles vaccines were also introduced. Three doses of killed virus vaccine elicited adequate MV-specific neutralizing and HI antibody, but little antibody to N and no antibody to F (CARTER et al. 1962; NORRBY and GOLLMAR 1975; ENDERS-RUCKLE 1967). Lack of antibody to F was apparently due to inactivation of the F protein during preparation of the vaccine (NORRBY and GOLLMAR 1975). Antibody responses were relatively short-lived, as was protection from infection (FULGINITI and ARTHUR 1969; RAUH and SCHMIDT 1965).

Cellular immune responses were also distinct from those induced by live attenuated virus since MV-specific DTH skin test responses and strong proliferative responses were observed (FULGINITI et al. 1968; NELSON et al. 1966; KRAUSE et al. 1978). This pattern suggests a preferential type 1 CD4 T cell response and suggests that the two types of vaccine (live and inactivated) stimulate T cell immunity in different ways. This could occur through differences in antigen presentation inherent in the processing of replicating and nonreplicating antigens or from the use of alum as an adjuvant.

The inactivated vaccine was in use for only a short period because of the appearance of an atypical form of measles in recipients of this vaccine exposed to wild-type virus in community outbreaks (CHERRY et al. 1972; BRODSKY 1972; RAUH and SCHMIDT 1965; NADER et al. 1968; MARTIN et al. 1979). Atypical measles was characterized by a high and prolonged fever, a more severe rash that began on the extremities and tended to become hemorrhagic or vesicular and a severe, often life threatening, pneumonitis. Serum contained high levels of antibody to the M protein (MACHAMER et al. 1980; NORRBY et al. 1981). Biopsies of the rash showed a combination of a DTH and Arthus-type reaction (ANNUNZIATO et al. 1982). Occasionally, severe local reactions were also seen in response to subsequent immunization with the live attenuated vaccine (FULGINITI et al. 1968; BUSER 1967; SCOTT and BONANNO 1967).

The pathogenesis of atypical measles remains unclear but may be due to lack of antibody to F allowing spread of virus, to lack of antibody in respiratory secretions (BELLANTI et al. 1969), to lack of a primed CD8 T cell response or to preferential stimulation of a type 1 rather than a type 2 CD4 T cell response

(KRAUSE et al. 1978). None of these mechanisms are mutually exclusive and it is possible that all contributed to the immunopathology observed as a part of atypical measles.

7 Immune Responses During Persistent Infection

Although there is no evidence that MV routinely causes persistent infection, it is the cause of two rare chronic progressive infections of the central nervous system (CNS), measles inclusion body encephalitis (MIBE) and subacute sclerosing panencephalitis (SSPE). MIBE is one manifestation of the progressive MV infection that occurs in immunosuppressed individuals (MUSTAFA et al. 1993). Progressive giant cell pneumonia is the most common manifestation (MITUS et al. 1959). Cellular immune responses are deficient and virus clearance is not effective. Often, these patients present without a rash (ENDERS et al. 1959; KAPLAN et al. 1992; MARKOWITZ et al. 1988), since generation of the rash requires an immune response to the virus. MIBE and giant cell pneumonia occur within weeks to months after infection.

In contrast, SSPE is a late complication of measles most common in individuals with apparently normal immune function and primary measles infection at a young age. SSPE occurs many years after infection. Although SSPE viruses have a variety of alterations in sequence compared to wild-type viruses there is no epidemiologic evidence to suggest that SSPE is caused by infection with a unique strain of MV. Instead, mutations probably accumulate during the extensive period of subclinical virus replication.

The determinants of host susceptibility to SSPE are unclear, but probably involve rare primary infection of the CNS and inadequate initial clearance of virus, perhaps associated with young age or the modulating influence of maternal antibody (RAMMOHAN et al. 1983). Shortly after the disease was recognized to be caused by MV a variety of subtle abnormalities of immune function were reported. However, there is no evidence that a particular inherited or acquired immune deficiency predisposes to SSPE. Chronic disease, associated with persistent infection may lead to these immunologic abnormalities.

Studies of the immune responses to MV in SSPE have shown very high titers of antiviral antibody. The protein specificities of the antibodies are similar to those seen in uncomplicated measles, including relatively low levels of antibody to the M protein (HALL et al. 1979). Isotypes produced are also similar to uncomplicated disease but the tendency to produce IgG4 is exaggerated (MATHIESEN et al. 1990). Elevated levels of antibody to MV in cerebrospinal fluid (CSF) are the result of intrathecal synthesis of immunoglobulin and are diagnostic of the disease (HADDAD et al. 1977; CONNOLLY et al. 1967; TOURTELLOTTE et al. 1981). The isotypes of MV-specific antibodies found in CSF include IgG, IgA and IgD (ESIRI et al. 1982; Patrick et al. 1990). Restricted populations of antibody-secreting B cells become resident in the CNS leading to the appearance of oligoclonal bands in the CSF

(Vandvik and Norrby 1973). Much of the locally produced antibody is complexed to MV antigens in the brain (Sotrel et al. 1983; Weil et al. 1975).

Cellular immune responses are indicated by the infiltration of mononuclear cells into areas of CNS virus infection. Most of these infiltrating cells are B cells and CD4 T cells and local production of TNF-α and IFN-γ is detectable (Nagano et al. 1991; Hofman et al. 1991). The CSF has increased levels of β2 microglobulin, soluble IL-2 receptor and soluble CD8 indicative of a local cellular immune reaction (Mehta et al. 1992). Studies of the responses of PBMCs to MV have shown MV-specific proliferative responses that are similar to those in individuals who recovered from infection (Brajczewska-Fischer et al. 1989). Large amounts of IL-1 are released spontaneously in culture. However, generation of MV-specific cytotoxic T cells is lower than in normal seropositive individuals suggesting that cytotoxic responses may be suppressed in chronic infection (Dhib-Jalbut et al. 1989).

8 Summary

The characteristic disease features of measles—fever and rash—are associated with the immune response to infection and are coincident with virus clearance. MV-specific antibody and CD4 and CD8 T cell responses are generated and contribute to virus clearance and protection from reinfection. During this same phase of immune activation immunologic abnormalities are also apparent. There is a generalized suppression of cellular immune responses that may contribute to increased susceptibility to other infections. Autoimmune disease may appear in the form of acute disseminated encephalomyelitis. If virus-specific immune responses are inadequate infection may progress with pulmonary or CNS manifestations, but without a rash. The pathogenesis of the rare disease SSPE, that occurs many years after primary infection is not clear, but immune responses show increased antibody to measles and cellular immune responses similar to those seen after uncomplicated infection.

References

Aicardi J, Goutieres F et al. (1977) Acute measles encephalitis in children with immunosuppression. Pediatrics 59: 232–239

Albrecht P, Ennis FA et al. (1977) Persistence of maternal antibody in infants beyond 12 months: mechanism of measles vaccine failure. J Pediatr 91: 715–178

Annunziato D, Kaplan MH et al. (1982) Atypical measles syndrome: pathologic and serologic findings. Pediatrics 70: 203–209

Arneborn P, Biberfeld G (1983) T lymphocyte subpopulations in relation to immunosuppression in measles and varicella. Infect Immun 39: 29–37

Barrett PN, Koschel K et al. (1985) Effect of measles virus antibodies on a measles SSPE virus persistently infected C6 rat glioma cell line. J Gen Virol 66: 1411–1421

Bech V (1959) Studies on the development of complement fixing antibodies in measles patients. J Immunol 83: 267–275

Beckford AP, Kaschula ROC et al. (1985) Factors associated with fatal cases of measles: A retrospective autopsy study. S Afr Med J 68: 858–863
Bellanti JA, Sanga RL et al. (1969) Antibody responses in serum and nasal secretions of children immunized with inactivated and attenuated measles-virus vaccines. N Engl J Med 280: 628–633
Bellini WJ, McFarlin DE et al. (1981) Immune reactivity of purified hemagglutinin of measles virus. Infect Immun 32: 1051–1057
Black FL (1989) Measles active and passive immunity in a worldwide perspective. Prog Med Virol 36: 1–33
Brajczewska-Fischer W, Iwinska B et al. (1989) Interleukin 1 and 2 production by peripheral blood mononuclear cells in subacute sclerosing panencephalitis and exacerabation of multiple sclerosis. Acta Neurol Scand 80: 390–393
Brodsky AL (1972) Atypical measles: severe illness in recipients of killed measles virus vaccine upon exposure to natural infection. J Am Med Assoc 222: 1415–1416
Buckland R, Giraudon P et al. (1989) Expression of measles virus nucleoprotein in Escherichia coli: use of deletion mutants to locate the antigenic sites. J Gen Virol 70: 435–441
Buser F (1967) Side reaction to measles vaccination suggesting the arthus phenomenon. N Engl J Med 277: 250–251
Carter CH, Conway TJ et al. (1962) Serologic response of children to inactivated measles vaccine. JAMA 179: 848–853
Carter MJ, Willcocks MM et al. (1982) Relationships between monoclonal antibody-binding sites on the measles virus haemagglutinin. J Gen Virol 63: 113–120
Cherry JD, Feigin RD et al. (1972) Atypical measles in children previously immunized with attenuated measles virus vaccines. Pediatrics 50: 712–717
Connolly JH, Allen IV et al. (1967) Measles virus antibody and antigen in subacute sclerosing panencephalitis. Lancet 1: 542–544
Coovadia HM, Wesley A et al. (1981) Measles, histocompatibility leukocyte antigen polymorphism, and natural selection in humans. J Infect Dis 144: 142–147
Crespi M, Struthers JK et al. (1988) Interferon status after measles virus infection. S Afr Med J 73: 711–712
Dhib-Jalbut S, Jacobson S et al. (1989) Impaired human leukocyte antigen-restricted measles virus-specific cytotoxic T-cell response in subacute sclerosing panencephalitis. Ann Neurol 25: 272–280
Donaldson SL, Kosco MH et al. (1986) Localization of antibody-forming cells in draining lymphoid organs during long-term maintenance of the antibody response. J Leukoc Biol 40: 147–157
Ehrnst AC (1975) Characterization of measles virus-specific cytotoxic antibodies by use of a chronically infected cell line. J Immunol 114: 1077–1082
Ehrnst A (1978) Separate pathways of C activation by measles virus cytotoxic antibodies: subclass analysis and capacity of F (ab) molecules to activate C via the alternative pathway. J Immunol 121: 1206–1212
Enders JF, McCarthy K et al. (1959) Isolation of measles virus at autopsy in cases of giant cell pneumonia without rash. N Engl J Med 261: 875–881
Enders-Ruckle G (1967) Some characteristics of immunity following natural measles and various forms of immunization. Arch Gesamte Virusforsch 22: 23–34
Esiri MM, Oppenheimer DR et al. (1982) Distribution of measles antigen and immunoglobulin containing cells in the CNS in subacute sclerosing panencephalitis (SSPE) and atypical measles encephalitis. J Neurol Sci 53: 29–43
Esolen LM, Ward BJ et al. (1993) Infection of monocytes during measles. J Infect Dis 168: 47–52
Fireman P, Friday G et al. (1969) Effect of measles virus vaccine on immunologic responsiveness. Pediatrics 43: 264–272
Fujinami RS, Oldstone MBA (1979) Antiviral antibody reacting on the plasma membrane alters measles virus expression inside the cell. Nature 279: 529–530
Fulginiti VA, Arthur JH (1969) Altered reactivity to measles virus. J Pediatr 75: 609–616
Fulginiti VA, Arthur JH et al. (1968) Altered reactivity to measles virus local reactions following attenuated measles virus immunization in children who previously received a combination of inactivated and attenuated vaccines. Am J Dis Child 115: 67–72
Furukawa S, Matsubara T et al. (1992) Kawasaki disease differs from anaphylactoid purpura and measles with regard to tumour necrosis factor-alpha and interleukin 6 in serum. Eur J Pediatr 151: 44–47
Garenne M, Leroy O et al. (1991) Child mortality after high-titre measles vaccines: prospective study in Senegal. Lancet 338: 903–907

Gendelman H, Wolinsky JS et al. (1984) Measles encephalitis: Lack of evidence of viral invasion of the central nervous system and quantitative study of the nature of demyelination. Ann Neurol 15: 353–360
Goretsky M (1990) Acute fatal leukoencephalopathy after interleukin-2 therapy. N Engl J Med 323: 1146–1147
Graves M, Griffin DE et al. (1984) Development of antibody to measles virus polypeptides during complicated and uncomplicated measles virus infections. J Virol 49: 409–412
Gray D, Skarvall H (1988) B-cell memory is short-lived in the absence of antigen. Nature 336: 70–73
Greenstein JI, McFarland HF (1983) Response of human lymphocytes to measles virus after natural infection. Infect Immun 40: 198–204
Griffin DE, Ward BJ (1993) Differential CD4 T cell activation in measles. J Infect Dis 168: 275–281
Griffin DE, Cooper SJ et al. (1985) Changes in plasma IgE levels during complicated and uncomplicated measles virus infections. J Allergy Clin Immunol 76: 206–213
Griffin DE, Moench TR et al. (1986) Peripheral blood mononuclear cells during natural measles virus infection: cell surface phenotypes and evidence for activation. Clin Immunol Immunopathol 40: 305–312
Griffin DE, Johnson RT et al. (1987) In vitro studies of the role of monocytes in the immunosuppression associated with natural measles virus infections. Clin Immunol Immunopathol 45: 375–383
Griffin DE, Moench TR et al. (1986) Peripheral blood mononuclear cells during natural measles virus infection: cell surface phenotypes and evidence for activation. Clin Immunol Immunopathol 40: 305–312
Griffin DE, Ward BJ et al. (1989) Immune activation during measles. N Engl J Med 320: 1667–1672
Griffin DE, Ward BJ et al. (1990a) Immune activation during measles: Interferon-gamma and neopterin in plasma and cerebrospinal fluid in complicated and uncomplicated disease J Infect Dis 161: 449–453
Griffin DE, Ward BJ et al. (1990b) Natural killer cell activity during measles. Clin Exp Immunol 81: 218–224
Griffin DE, Ward BJ et al. (1992) Immune activation during measles: Beta-2-microglobulin in plasma and cerebrospinal fluid in complicated and uncomplicated disease. J Infect Dis 166: 1170–1173
Haddad FS, Risk WS et al. (1977) Subacute sclerosing panencephalitis in the Middle East: report of 99 cases. Ann Neurol 1: 211–217
Hall WW, Lamb RA et al. (1979) Measles and subacute sclerosing panencephalitis virus proteins: lack of antibodies to the M protein in patients with subacute sclerosing panencephalitis. Proc Natl Acad Sci USA 76: 2047–2051
Halsey NA, Boulos R et al. (1985) Response to measles vaccine in Haitian infants 6 to 12 months old. N Engl J Med 313: 544–549
Hirsch RL, Mokhtarian F et al. (1981) Measles virus vaccination of measles seropositive individuals suppresses lymphocyte proliferation and chemotactic factor production. Clin Immunol Immunopathol 21: 341–350
Hirsch RL, Griffin DE et al. (1984) Cellular immune responses during complicated and uncomplicated measles virus infections of man. Clin Immunol Immunopathol 31: 1–12
Hofman FM, Hinton DR et al. (1991) Lymphokines and immunoregulatory molecules in subacute sclerosing panencephalitis. Clin Immunol Immunopathol 58: 331–342
Ilonen J, Makela MJ et al. (1990) Cloning of human T cells specific for measles virus haemagglutinin and nucleocapsid. Clin Exp Immunol 81: 212–217
Jacobson S, Richert JR et al. (1984) Measles virus-specific T4+ human cytotoxic T cell clones are restricted by class II HLA antigens. J Immunol 133: 754–757
Jacobson S, Sekaly RP et al. (1989) HLA class II-restricted presentation of cytoplasmic measles virus antigens to cytotoxic T cells. J Virol 63: 1756–1762
Joffe MI, Rabson AR (1981) Defective helper factor (LMC) production in patients with acute measles infection. Clin Immunol Immunopathol 20: 215–223
Johnson RT, Griffin DE et al. (1984) Measles encephalomyelitis—clinical and immunologic studies. N Engl J Med 310: 137–141
Joseph BS, Oldstone BA (1975) Immunologic injury in measles virus infection II. Suppression of immune injury through antigenic modulation. J Exp Med 142: 864–876
Joseph BS, Lampert PW et al. (1975) Replication and persistence of measles virus in defined subpopulations of human leukocytes. J Virol 16: 1638–1649
Kaplan LJ, Daum RS et al. (1992) Severe measles in immunocompromised patients. JAMA 267: 1237–1241
King JC Jr, Lichnestein R et al. (1993) Measles, mumps, and rubella antibodies in vaccinated Baltimore children. Amer J Dis Child 147: 558–560

Krause PJ, Cherry JD et al. (1978) Revaccination of previous recipients of killed measles vaccine: clinical and immunologic studies. J Pediatr 93: 565–571
Kreth HW, ter Meulen V et al. (1979) Demonstration of HLA restricted killer cells in patients with acute measles. Med Microbiol Immunol 165: 203–214
Larner AC, Petricoin EF et al. (1993) IL-4 attenuates the transcriptional activation of both IFN-alpha and IFN-gamma-induced cellular gene expression in monocytes and monocytic cell lines. J Immunol 150: 1944–1950
Leopardi R, Ilonen J et al. (1993) Effect of measles virus infection on MHC class II expression and antigen presentation in human monocytes. Cell Immunol 147: 388–396
Leopardi R, Vainionpaa R et al. (1992) Measles virus infection enhances IL-1 beta but reduces tumor necrosis factor-alpha expression in human monocytes. J Immunol 149: 2397–2401
Litvak AM, Sands IJ et al. (1943) Encephalitis complicating measles: report of fifty-six cases with followup studies in thirty-two. Am J Dis Child 65: 265–295
Lucas CJ, Biddison WE et al. (1982) Killing of measles virus-infected cells by human cytotoxic T cells. Infect Immun 38: 226–232
Lucas J (1790) An account of uncommon symptoms succeeding the measles; with additional remarks on the infection of measles and smallpox. Lond Med J 11: 325–331
Machamer CE, Hayes EC et al. (1980) Antibodies against the measles matrix polypeptide after clinical infection and vaccination. Infect Immun 27: 817–825
Malvoisin E, Wild F (1990) Contribution of measles virus fusion protein in protective immunity: anti-F monoclonal antibodies neutralize virus infectivity and protect mice against challenge. J Virol 64: 5160–5162
Mandalenaki-Asfi C, Liakopoulou P et al. (1976) Rosette-forming lymphocytes and measles vaccination. J Pediatr 88: 74–75
Markowitz LE, Chandler FW et al. (1988) Fatal measles pneumonia without rash in a child with AIDS. J Infect Dis 158: 480–483
Markowitz LE, Preblud SR et al. (1990) Duration of live measles vaccine-induced immunity. Pediatr Infect Dis J 9: 101–110
Martin DB, Weiner LB et al. (1979) Atypical measles in adolescents and young adults. Ann Intern Med 90: 877–881
Martinez OM, Gibbons RS et al. (1990) IL-4 inhibits IL-2 receptor expression and IL-2 dependent proliferation of human T cells. J Immunol 144: 2211–2215
Mathiesen T, Hammarstrom L et al. (1990) Aberrant IgG subclass distribution to measles in healthy seropositive individuals, in patients with SSPE and in immunoglobulin-deficient patients. Clin Exp Immunol 80: 202–205
Mehta PD, Kulczycki J et al. (1992) Increased levels of beta-2-microglobulin, soluble interleukin-2 receptor, and soluble CD8 in patients with subacute sclerosing panencephalitis. Clin Immunol Immunopathol 65: 53–59
Merz DC, Scheid A et al. (1980) Importance of antibodies to the fusion glycoprotein of paramyxoviruses in the prevention of spread of infection. J Exp Med 151: 275–288
Miller DL (1964) Frequency of complications of measles. 1963 Br Med J 2: 75–78
Mitus A, Enders JF et al. (1959) Persistence of measles virus and depression of antibody formation in patients with giant-cell pneumonia after measles. N Engl J Med 261: 882–889
Moench TR, Griffin DE et al. (1988) Acute measles in patients with and without neurological involvement: distribution of measles virus antigen and RNA. J Infect Dis 158: 433–442
Morley D (1969) Severe measles in the tropics. Br Med J 1: 297–300
Mosmann TR, Coffman RL (1987) Two types of mouse helper T cell clone: implications for immune regulation. Immunol Today 8: 223–227
Mustafa MM, Weitman SD et al. (1993) Subacute measles encephalitis in the young immunocompromised host: report of two cases diagnosed by polymerase chain reaction and treated with ribavirin and review of the literature. Clin Infect Dis 16: 654–660
Nader PR, Horwitz MS et al. (1968) Atypical exanthem following exposure to natural measles: eleven cases in children previously inoculated with killed vaccine. J Pediatr 72: 22–28
Nagano I, Nakamura S et al. (1991) Immunocytochemical analysis of the cellular infiltrate in brain lesions in subcute sclerosing panencephalitis. Neurology 41: 1639–1642
Nelson JD, Sandusky G et al. (1966) Measles skin test and serologic response to intradermal measles antigen. JAMA 198: 185–186
Norrby E, Enders-Ruckle G et al. (1975) Differences in the appearance of antibodies to structural components of measles virus after immunization with inactivated and live virus. J Infect Dis 132: 262
Norrby E, Göllmar Y (1972) Appearance and persistence of antibodies against different virus components after regular measles infections Infect. Immun 6: 240–247

Norrby E, Gollmar Y (1975) Identification of measles virus-specific hemolysis-inhibiting antibodies separate from hemagglutination-inhibiting antibodies. Infect Immun 11: 231–239

Norrby E, Orvell C et al. (1981) Antibodies against measles virus polypeptides in different disease conditions. Infect Immun 34: 718–724

Panum P (1938) Observations made during the epidemic of measles on the Faroe Islands in the year 1846. Med Class 3: 829–886

Patrick BA, Mehta PD et al. (1990) Measles virus-specific immunoglobulin D antibody in cerebrospinal fluid and serum from patients with subacute sclerosing panencephalitis and multiple sclerosis. J Neuroimmunol 26: 69–74

Rammohan KW, McFarland HF et al. (1983) Antibody-mediated modification of encephalitis induced by hamster neurotropic measles virus. J Infect Dis 147: 546–550

Rauh LW, Schmidt R (1965) Measles immunization with killed virus vaccine. Am J Dis Child 109: 232–237

Rose JW, Bellini WJ et al. (1984) Human cellular immune response to measles virus polypeptides. J Virol 49: 988–991

Rota JS, Hummel KB et al. (1992) Genetic variability of the glycoprotein genes of current wild-type measles isolates. Virology 188: 135–142

Ruckle G, Rogers KD (1957) Studies with measles virus. II. Isolation of virus and immunologic studies in persons who have had the natural disease. J Immunol 78: 341–355

Salonen R, Ilonen J et al. (1988) Measles virus infection of unstimulated blood mononuclear cells in vitro: antigen expression and virus production preferentially in monocytes. Clin Exp Immunol 71: 224–228

Schluederberg A (1965) Immune globulins in human viral infections. Nature 205: 1232–1233

Schneider-Schaulies S, Liebert UG et al. (1992) Antibody-dependent transcriptional regulation of measles virus in persistently infected neural cells. J Virol 66: 5534–5541

Scott TFM, Bonanno DE (1967) Reactions to live-measles-virus vaccine in children previously inoculated with killed-virus vaccine. N Engl J Med 277: 248–250

Sheshberadaran H, Norrby E (1986) Characterization of epitopes on the measles virus hemagglutinin. Virology 152: 58–65

Shiozawa S, Yoshikawa N et al. (1988) A sensitive radioimmunoassay for circulating alpha-interferon in the plasma of healthy children and patients with measles virus infection. Clin Exp Immunol 73: 366–369

Siber GR, Werner BG et al. (1993) Interference of immune globulin with measles and rubella immunization. J Pediatr 122: 204–211

Sotrel A, Rosen S et al. (1983) Subacute sclerosing panencephalitis: an immune complex disease? Neurology 33: 885–890

Stephenson JR, ter Meulen V (1979) Antigenic relationships between measles and canine distemper viruses: comparison of immune response in animals and humans to individual virus-specific polypeptides. Proc Natl Acad Sci USA 76: 6601–6605

Sullivan JL, Barry DW et al. (1975) Measles infection of human mononuclear cells. I. Acute infection of peripheral blood lymphocytes and monocytes. J Exp Med 142: 773–784

Tamashiro VG, Perez HH et al. (1987) Prospective study of the magnitude and duration of changes in tuberculin reactivity during complicated and uncomplicated measles. Pediatr Infect Dis J 6: 451–454

Taylor MJ, Godfrey E et al. (1991) Identification of several different lineages of measles virus. J Gen Virol 72: 83–88

ter Meulen V, Loffler S et al. (1981) Antigenic characterization of measles and SSPE virus haemagglutinin by monoclonal antibodies. J Gen Virol 67: 357–364

Tew JG, Kosco MH et al. (1990) Follicular dendritic cells as accessory cells. Immunol Rev 117: 185–211

Tourtellotte WW, Ma BI et al. (1981) Quantification of de novo central nervous system IgG measles antibody synthesis in SSPE. Ann Neurol 9: 551–556

Trudgett A, Gould EA et al. (1981) Antigenic difference in the hemagglutinin of measles and related viruses. Virology 109: 180–182

van Binnendijk R, Poelen MCM et al. (1989) Measles virus-specific human T cell clones. Characterization of specificity and function of CD4 helper/cytotoxic and $CD8^{+}$ cytotoxic T cell clones. J Immunol 142: 2847–2854

van Binnendijk RS, Poelen MCM et al. (1990) The predominance of $CD8^{+}$ T cells after infection with measles virus suggests a role for $CD8^{+}$ class I MHC-restricted cytotoxic T lymphocytes (CTL) in recovery from measles. J Immunol 144: 2394–2399

van Binnendijk RS, van Baalen CA et al. (1992) Measles virus transmembrane fusion protein systhesized de novo or presented in immunostimulating complexes is endogenously processed for HLA class I- and class II- restricted cytotoxic T cell recognition. J Exp Med 176: 119–128

van Binnendijk RS, Versteeg-van Oosten JPM et al. (1993) Human HLA class I- and HLA class II-restricted cloned cytotoxic T lymphocytes identify a cluster of epitopes on the measles virus fusion protein. J Virol 67: 2276–2284
Vandvik B, Norrby E (1973) Oligoclonal IgG antibody response in the central nervous system to different measles virus antigens in subacute sclerosing panencephalitis. Proc Natl Acad Sci USA 70: 1060–1063
Von Pirquet C (1908) Verhalten der kutanen Tuberkulinreaktion wahrend der Masern. Dtsch Med Wochenschr 34: 1297–1300
Ward BJ, Griffin DE (1993) Changes in cytokine production after measles virus vaccination: predominant production of IL-4 suggests induction of a Th2 response. Clin Immunol Immunopathol 67: 171–177
Ward BJ, Johnson RT et al. (1990) Spontaneous proliferation of peripheral mononuclear cells in natural measles virus infection: identification of dividing cells and correlation with mitogen responsiveness. Clin Immunol Immunopathol 55: 315–326
Ward BJ, Johnson RT et al. (1991) Cytokine production in vitro and the lymphoproliferative defect of natural measles virus infection. Clin Immunol Immunopathol 61: 236–248
Weil ML, Leiva WA et al. (1975) Release of bound immunoglobulin from SSPE brain by acid elution. J Immunol 115: 1603–1606
Wiesmuller K-H, Handtmann D et al. (1992) Heterogeneity of linear B cell epitopes of the measles virus fusion protein reacting with late convalescent sera. J Gen Virol 73: 2211–2216
Wild TF, Malvoisin E et al. (1991) Measles virus: both the haemagglutinin and fusion glycoproteins are required for fusion. J Gen Virol 72: 439–442
Wu VH, McFarland H et al. (1993) Measles virus-specific cellular immunity in patients with vaccine failure. J Clin Microbiol 31: 118–122
Zweiman B, Pappagianis D et al. (1971) Effect of measles immunization on tuberculin hypersensitivity and in vitro lymphocyte reactivity. Int Arch Allergy Appl Immunol 40: 834–841

Monkeys in Measles Research

R.S. van Binnendijk, R.W.J. van der Heijden, and A.D.M.E. Osterhaus

1 Introduction

The identification of measles virus (MV) as the causative agent of measles was first described in 1911, when filtered respiratory tract secretions of measles patients were inoculated into macaque monkeys causing measles-like symptoms in these animals (Goldberger and Anderson 1911). It was not until 1954 that the virus could be isolated and adapted to growth in vitro in several cell lines of primate and nonprimate origin (Enders and Peebles 1954; Ruckle and Rogers 1957; Katz et al. 1958; Enders 1962). This provided the basis for extensive biological research on the pathogenicity of MV and led to diagnostic methods for measles and eventually to the development of measles vaccines (Katz and Enders 1959; Enders et al. 1960; Katz 1965). Apart from humans, nonhuman primates proved susceptible animal species for MV infection, either by contracting the infection from humans during captivity or by experimental infection with different clinical specimens. Although marmosets, macaques and several other monkey species have been used in studies concerning the host range and the virulence of different

Erasmus University, Department of Virology, P.O. Box 1738, 3000 DR Rotterdam, The Netherlands

strains of MV, macaque species have been studied most extensively. It has been shown that the pathogenesis of MV infection in macaques is similar to that of measles in humans.

Besides different primate species, also rodents (hamster, rat, mice) can be infected experimentally with MV. However, replication of MV in rodents is largely restricted to neuroadapted strains of the virus and to the central nervous system (CNS) of the animals (INAGAWA and ADAMS 1958; WAKSMAN et al. 1962; BURNSTEIN et al. 1964; MATUMOTO et al. 1964; JANDA et al. 1971; GRIFFIN et al. 1974; HERNDON et al. 1975, NEIGHBOUR et al. 1978; CHAN 1985; CARRIGAN 1986; CARRIGAN and KABAKOFF 1987; OHUCHI et al. 1984,1986; LIEBERT and TER MEULEN 1987; LIEBERT et al. 1988). Therefore, the rodent model may be of special interest to study the pathogenesis of infections with different MV strains, the regulation of viral gene expression, and the contribution of B and T cell mediated immune responses to protection from CNS infection (DRILLIEN et al. 1988; DE VRIES et al. 1988; MALVOISIN and WILD 1990; BANKAMP et al. 1991; BRINCKMANN et al. 1991; NIEWIESK et al. 1993). The selective replication of MV in the CNS also implicates the limitations of the rodent models, since the CNS is rarely involved in clinical manifestations of MV infection of humans.

One of the priorities in MV research is the development of a suitable animal model which may provide more insight into the pathogenesis of measles, the induction of humoral and cellular immune responses and molecular correlates of virus attenuation. Moreover, a suitable animal model will offer the opportunity to evaluate the efficacy and safety of novel generations of measles vaccines and novel vaccination strategies. In this chapter, we review the present knowledge concerning the pathogenesis of MV infection and immune defense mechanisms underlying the disease in monkeys.

2 Pathogenicity of Measles Virus in Monkeys

2.1 Introduction

Following the most early transfer experiments of measles from humans to monkeys (JOSIAS 1898; GOLDBERGER and ANDERSON 1911), blood or throat washings obtained from measles patients were repeatedly transferred to macaque monkeys to confirm the etiology of the disease and to establish a model for measles. Though the response was extremely variable and only observed in part of the inoculated animals, the most frequently occurring clinical signs reported were exanthema and fever (LUCAS and PRIZER 1912; BLAKE and TRASK 1921; DEGWITZ 1927; KRAFT 1932; PLOTZ 1938; HURST and COOKE 1941; SCHAFFER et al. 1941).

After the successful isolation and propagation of MV in vitro (ENDERS and PEEBLES 1954), inoculation experiments were continued in rhesus (*Macaca mulatta*), cynomolgus (*Macaca fascicularis*) and in baboon (*Papio hamadryas* and *P. hybridus*) monkeys. In these studies not only clinical specimens were used, but also defined strains of MV (PEEBLES et al. 1957; ENDERS et al. 1960; SERGIEV et al. 1960; YAMANOUCHI et al. 1970, 1973; ALBRECHT et al. 1972, 1977a; SAKAGUCHI et al. 1986; KOBUNE et al. 1990; VAN BINNENDIJK et al. 1994). In most of these experiments, clinical manifestations of measles were not observed or the observed symptoms were restricted to mild fever, sight leukopenia and mild respiratory symptoms, although symptoms more characteristic for measles such as conjunc-tivitis and maculopapular rashes have been reported occasionally (PEEBLES et al. 1957; SERGIEV et al. 1960; KOBUNE et al. 1990). However, the susceptibility of many monkey species to a variety of MV strains and the close similarity between measles pathogenesis in humans and macaque monkeys were unequivocally demonstrated by a variety of other observations. These included; (1) the presence of viral antigen and multinucleated giant cells of the reticular type in lymphoid organs and in epithelial tissues including the trachea and skin of killed animals (TANIGUCHI et al. 1954; SERGIEV et al. 1960; NII et al. 1964; ONO et al. 1970; YAMANOUCHI et al. 1970, 1973; HALL et al. 1971; SAKAGUCHI et al. 1986; KOBUNE et al. 1990); (2) the development of a disseminated MV infection in lung macrophages, epithelial cells of the pharynx and in peripheral blood mononuclear cells (PBMCs) during the first 2 weeks of infection (Peebles et al. 1957; HICKS et al. 1977; VAN BINNENDIJK et al. 1994); and (3) the appearance and persistence if virus-specific serum antibodies shortly after the peak of viremia (ENDERS et al. 1960; VAN BINNENDIJK et al. 1994).

2.2 Outbreaks of Measles in Monkey Colonies

Neither cynomolgus nor rhesus monkeys are infected with MV in their natural environment but they are often exposed to this infection when captured (PEEBLES et al. 1957; HABERMANN and WILLIAMS 1957; MEYER et al. 1962; POTKAY et al. 1966; SHISHIDO 1966; HALL et al. 1971; REMFRY 1976; WELSHMAN 1989). Previously it has been described that measles can spread rapidly through a captive population (MEYER et al. 1962; SHISHIDO 1966; WELSHMAN 1989). Only in part of the population could clinical signs be observed, in some cases associated with the development of a maculopapular rash and conjunctivitis (PEEBLES et al. 1957; Potkay et al. 1966; HALL et al. 1971; REMFRY 1976; WELSHMAN 1989). However, in all animals studied, the presence of multinucleated giant cells in skin lesions and in lymphoid organs and the appearance of MV-specific hemagglutination-inhibiting serum antibodies confirmed the etiology of the disease and the spread of virus through the whole population within 3 weeks of captivity (MEYER et al. 1962; SHISHIDO 1966; POTKAY et al. 1966; YAMANOUCHI et al. 1969; HALL et al. 1971; WELSHMAN 1989). Substantial evidence for the contagiousness of measles among monkeys came from experimentally infected marmosets (*Saguinus mystax* and *Saguinus labiatus*). Animals infected with two wild-type strains of MV (JM and Edmonston) were capable of rapidly spreading the virus to separately housed MV-seronegative control animals by aerosol, which is also the main mechanism of transmission in humans (LORENZ and ALBRECHT 1980).

Epizootics of measles in monkeys during captivity are not restricted to marmosets and macaques, Outbreaks of measles have been reported in colobus monkeys (*Colobus guereza*), in which the majority of pathological and clinical symptoms were similar to the disease in macaques, except for the absence of rashes and Koplik spots (HIME and KEYMER 1975). From these colobus monkeys virus could also be isolated and neutralized by MV-specific human and macaque antisera (SCOTT and KEYMER 1975). Other examples of nonhuman primate species susceptible to measles during captivity are spider monkeys (*Ateles Spp*), talapoins (*Cercopithecus talapoin*) and chimpanzees (*Pan troglodytes*). Although these latter species often lack the more typical clinical signs of the disease, the occurrence of MV infections have been recognized by the appearance of specific serum antibodies and the presence of multinucleated giant cells in the lungs of fatal cases (MACARTHUR et al. 1979).

2.3 Experimental Measles Virus Infections in Monkeys

PEEBLES et al. (1957) and ENDERS et al. (1960) were the first to study experimentally infected monkeys in detail. They infected nine MV-seronegative cynomolgus monkeys by the intranasal and intravenous route with the virulent MV Edmonston strain that had been passaged in human kidney cells. While only four animals developed maculopapular rashes, viremia could be detected in blood specimens of eight animals between days 7 and 11 postinoculation. Adaptation of the virulent MV Edmonston strain by passaging of the virus through chicken embryos in vivo and through chicken embryo cell cultures in vitro resulted in the first preparation of attenuated measles vaccines (MILOVANOVIC et al. 1957; KATZ et al. 1958). Infection of monkeys with these attenuated MV strains by either a combined intravenous-intranasal, a combined intracerebral-intracisternal or a subcutaneous route of inoculation appeared to be clinically silent and did not cause a detectable viremia. However, virus could be recovered from the nasal secretions of a number of animals that were infected by the subcutaneous route. Challenge experiments demonstrated that all immunized animals were protected against a subsequent combined intranasal-intravenous infection with the virulent MV Edmonston strain. Though local virus replication still occurred in the upper respiratory tract of these animals after challenge, clinical symptoms were absent and a systemic viremia was not observed (ENDERS et al. 1960). The initial infection experiments in a primate model provided a basis for attenuated MV strains to be used as safe and effective measles vaccines. Also, it was reassuring that the intracerebral inoculation of monkeys with attenuated strains of MV never elicited neuronal complications commonly associated with acute measles in humans (ENDERS et al. 1960; ALBRECHT et al. 1981).

Recently, we and others have infected cynomolgus monkeys with wild-type MVs isolated and propagated in Epstein-Barr virus (EBV)-transformed marmoset and human B lymphoblastoid cell lines (B-LCLs). These cell lines proved to be very sensitive indicator cells for the detection of MVs in clinical specimens of both humans and monkeys and in this way virulence for cynomolgus monkeys proved to be preserved (KOBUNE et al. 1990; VAN BINNENDIJK et al. 1994). Though the results of these experiments were in principal comparable to those published earlier by PEEBLES et al. (1957) and by ENDERS et al. (1960), monkeys showed a higher susceptibility to these wild-type MV isolates. A more extensive replication of the virus in PBMCs and lung macrophages and a more widely distributed giant cell formation in the lymphoid tissues were observed when compared to infections with other laboratory and attenuated strains of MV (YAMANOUCHI et al. 1970; KOBUNE et al. 1990; VAN BINNENDIJK et al. 1994). In contrast to the findings of PEEBLES et al. (1957), we were also able to detect a measles viremia in monkeys immunized with standard doses of an attenuated (Schwartz) measles vaccine. This may be related to the use of a more sensitive detection based on human B-LCLs rather than other continuous cell lines for virus isolation (VAN BINNENDIJK et al. 1994). Therefore it may be speculated that the virulence of wild-type MVs and their pathogenesis in monkeys may parallel the human situation, but may be strongly dependent on the strain and in vitro passage history of the MV strain used.

2.4 Encephalitogenicity of Measles Virus in Marmosets and Macaques

Whereas measles in macaques can be depicted as a rather mild disease with a normal measles-like pathogenesis, aberrant and fatal measles has been reported in a colony of marmosets, causing hundreds of animals to die within a period of several months. The disease was caused by a MV strain which could be isolated from moribund animals. Many of the animals showed drooping, puffy upper eyelids and mucous nasal discharge prior to a progressive lethargy. Typical measles skin reactions were almost never part of the symptoms. The most characteristic histopathological finding was an interstitial pneumonitis with giant cell formation (LEVY and MIRKOVIC 1971). LORENZ and ALBRECHT (1980) and ALBRECHT et al. (1980, 1981) corroborated these findings, showing widespread MV infections in the respiratory, gastrointestinal and other visceral tissues of marmosets of the *Saguinus mystax* subspecies when these animals were experimentally infected with a wild-type strain of MV (JM) by the respiratory route. Clinical signs and gross pathological findings lacked the characteristic features of measles and most of the infected animals died within 2 weeks due to a widespread gastroenterocolitis (ALBRECHT et al. 1980). As animals previously vaccinated with inactivated measles vaccines not only survived these infections but developed high titers of MV-specific antibody as well (LORENZ and ALBRECHT 1980; ALBRECHT et al. 1980), a primary deficiency in lymphocyte function has been implicated in what are probably the most virulent MV infections reported in monkeys.

The CNS of monkeys is rarely involved after parenterally administered MVs (ENDERS et al. 1960; ALBRECHT et al. 1980). However, marmosets infected intracerebrally with either of two wild-type strains of MV (JM and Edmonston), but not with attenuated measles vaccine, developed encephalitis in the absence of other clinical symptoms (Albrecht et al. 1981). The Edmonston strain was found to be considerably more neurotropic in these animals than the JM strain, causing perivascular cuffs, proliferation of glial cells, multinucleated giant cells and inflammatory response in brain tissues. Most of the animals inoculated with the JM strain showed a milder brain disease but died of the visceral form of measles infection (Albrecht et al. 1980) as was also reported after the parenteral administration of the JM strain (Albrecht et al. 1981). Marmosets thus seem to be a very sensitive indicator of the viscerotropic and neurotropic properties of wild-type and attenuated MVs. However, due to the lack of characteristic features of measles and the postulated immune dysfunction, the use of these animals seems less satisfactory for studies concerning MV pathogenesis and immune responses involved in recovery and protection from the natural disease.

Similar infection experiments carried out with the same wild-type strains and attenuated strains of MV in macaque monkeys did not result in measles encephalitis (Enders et al. 1960; Albrecht et al. 1972). However, intracerebral inoculation of rhesus monkeys with a hamster brain-adapted subacute sclerosing panencephalitis (SSPE) strain resulted in a chronic, progressive encephalitis, resembling characteristic features of SSPE in humans (Albrecht et al. 1972, 1977a). Interestingly, the disease ran a subacute or chronic course only in animals with preexisting immunity and was most vigorously induced with the virus passaged through monkey brain. Clinical disease was characterized by ataxia, lethargy and weakness prior to coma, and accompanied by extremely high antibody titers in serum and cerebrospinal fluid which were comparable to those found in SSPE. The striking similarities of the brain disease in monkeys with SSPE may make primates suitable model for studies concerning the mechanisms of SSPE and possible therapeutic approaches for this disease (Albrecht et al. 1977a).

3 Measles Virus-Specific Immunity in Monkeys

3.1 Introduction

Recovery of humans from measles and the subsequent protection against reinfection in humans may depend on virus-neutralizing antibody and virus-specific cell mediated immune responses. Shortly after the onset of the maculopapular rash, which is a hallmark for the development of cell-mediated immune response, MV-neutralizing antibody can be detected in the circulation. The appearance of neutralizing serum antibodies and the protective effect of passively transferred MV-specific antibody preparations to children for measles prophylaxis, demonstrates the importance of antibody-mediated immune response in the protection from MV infection (reviewed by Norrby and Oxman 1990). Passive transfer of polyclonal or monoclonal MV-neutralizing antibodies directed against either of the two MV glycoproteins, the hemagglutinin (H)and fusion (F) proteins proved to be effective in protecting naive mice and Lewis rats against MV-induced encephalitis (GIRAUDON and WILD 1985; SATO et al. 1989; MALVOISIN and WILD 1990). However, most agammaglobulinemic children seem to develop the normal clinical symptoms after MV infection and to develop protective immunity to reinfection. In contrast, children with functional T cell deficiencies often develop a fatal giant cell pneumonia after infection with MV (GOOD and ZAK 1956; MITUS et al. 1959; BURNET 1968). These observations suggest that cell-mediated immunity, probably in combination with neutralizing antibody, plays a major role in the control of MV infections rather than a unique role for antibody-mediated immune response. Although the transient immunosuppression associated with measles (reviewed by MCCHESNEY and OLDSTONE 1989) has initially hampered the investigation of MV-specific cell mediated immunity, recent studies have evaluated the kinetics and role of MV-specific cellular cytotoxic T lymphocytes (CTLs) (reviewed by UYTDEHAAG et al. 1993). From these studies, it was concluded that CTLs of the CD8 rather than of the CD4 phenotype are involved in the clearance of virus infected cells and in recovery from measles, T cells of both phenotypes may function as memory cells in maintaining life-long protection against measles in immune individuals (KRETH et al. 1979; LUCAS et al. 1982; JACOBSON et al. 1984,1987; VAN BINNENDIJK et al. 1989, 1990). It has been shown that passive transfer of MV-specific $CD4^{+}$ and $CD8^{+}$ T cell lines to naive mice and Lewis rats protect them from MV-induced encephalitis (BANKAMP et al. 1991; BRINCKMANN et al. 1991). It is not clear, whether the transferred T cell directly functioned as CTLs in these models. NIEWIESK et al. (1993) have clearly demonstrated that the occurrence of an MV-induced encephalitis correlated with an inefficient induction of MV-specific $CD8^{+}$ T Cells.

3.2 The Immunosuppressive Effects of Measles in Monkeys

MV infection in human is known to exert a transient immunosuppression. It has been suggested that the suppression of T cell mediated lymphoproliferation and antibody production by B lymphocytes to a number of stimulating mitogens and antigens are a direct result of virus replication in these cells during the acute stage of the disease (reviewed by MCCHESNEY and OLDSTONE 1989). An as yet unknown autocrine mechanism may be responsible for the observed inhibition, as the effector functions of these cells are dramatically impaired only when they are infected prior to their differentiation. During an outbreak of measles at a primate research center MV was detected in the PBMCs of infected rhesus monkeys. In these monkeys, pokeweed mitogen (PWM)-driven IgM and IgG synthesis was markedly reduced when compared to uninfected control animals. Immunoglobulin secretion was still reduced several months after infection but returned to normal levels within 1 year (MCCHESNEY et al. 1989). An enhanced mitogenic response was actually seen shortly after infection, when measles viremia and a marked leukopenia were also reported. Thus MV infection may not disseminate as rapidly through lymphocytes in vivo as it does under in vitro conditions and may rather activate certain populations of lymphocytes or even alter the circulatory pattern of lymphocytes shortly after infection (Hicks et al. 1977).

Recently, Contreras and Furesz (1992) described the immunosuppressive effects of MV infections on the outcome of the neurovirulence test for oral poliovirus vaccines in a colony of macaques in Canada. Although the origin and history of the MV infections were unknown, the prevalence and increase of high titers of specific serum antibody in these monkeys, which also paralleled the increase in the prevalence of measles in the population of Ontario, highly correlated with a more pronounced severity of poliomyelitis. Taken together, these observations indicate that the mechanisms of MV-induced immunosuppression in humans and macaque monkeys are similar.

3.3 Measles Virus Specific Annses in Macaques

As stated above, it is most likely that both virus-neutralizing antibody and virus-specific CTLs contribute to virus clearance after MV infection and to protective immunity in monkeys. Moreover, successful protection against MV infection induced by immunization with live attenuated measles vaccines may depend on the same immune mechanisms. However, further studies are necessary to define the specific contribution of individual components of the immune system. Since macaques develop clinical signs and a pathogenesis of MV infections similar to measles in humans, they may provide a suitable animal model to study protective immune responses elicited by either vaccination or MV infection and to evaluate the efficacy and safety of measles vaccines.

Epizootics of measles in captive populations of macaque monkeys were always accompanied by the appearance of MV-specific serum antibodies, closely resembling the kinetics of antibody responses after acute measles in humans. The persistence of specific serum antibodies in monkeys after infection may also be part of the protective immune response as no subsequent reinfections have been reported in the infected populations (Peebles et al. 1957; Meyer et al. 1962; Potkay et al. 1966; Shishido 1966; Hall et al. 1971; Remfry 1976; Welshman 1989).

Other investigators have reported similar types of MV-specific antibody responses in macaques after the administration of different strains of virulent MV, attenuated measles vaccines and inactivated MV preparations. Strong virus-neutralizing, complement-fixing and hemagglutination-inhibiting antibody responses were detected in the sera of monkeys 1-5 weeks after immunization, which declined to persistent levels over a period of 3 years (Peebles et al. 1957; Enders et al. 1960; Yamanouchi et al. 1969; Hicks and Albrecht 1976; MacArthur et al. 1982; de Vries et al. 1988; Welshman 1989; Van Binnendijk et al. 1994). In contrast to inactivated MV preparations, infectious MVs, especially when administered intranasally or intratracheally, often produced higher levels of virus-neutralizing serum antibodies in monkeys, which may be due to extensive virus replication and dissemination in the cells of the respiratory tract (Hicks et al. 1977; de Vries et al. 1988; van Binnendijk et al. 1994). MV-infected monkeys also developed antibodies mediating a complement-dependent lysis of MV-infected target cells (cytolytic antibodies) (PEEBLES et al. 1957; HICKS and ALBRECHT 1976). Immunization of monkeys with heat-inactivated MV resulted in very low concentrations or even the absence of cytolytic antibodies, which may be related to the general failure of such preparations to induce functional antibodies against the F glycoprotein of MV (HICKS and ALBRECHT 1976, reviewed by NORRBY 1985). However, the induction of functional and long-lived F protein-specific antibody responses has been reported in cynomolgus monkeys immunized with the F protein incorporated in an immune stimulating complex (ISCOM) (DE VRIES et al. 1988). These observations are of major importance as a severe atypical measles syndrome (AMS) has been reported in MV-infected children who had been vaccinated with inactivated measles vaccines 2-4 years before exposure to the virus. These vaccine failures were initially attributed to the inability of inactivated measles vaccines to induce the proper antibody response against the F protein of MV (reviewed by NORRBY 1985), although other mechanisms may be involved. These include an altered immunoglobulin subclass distribution of the antibody response (MATHIESEN et al. 1990) and the general failure of inactivated MV to induce a properly balanced T cell response (VAN BINNENDIJK et al. 1990, 1992).

3.4 Cell-Mediated Immune Responses in Macaques

Today, substantial knowledge is available concerning the role of antibody response induced by MV infection in humans and in monkeys; however parameters of T cell-mediated immune responses have only been studied extensively in humans and rodents (reviewed by UYTDEHAAG et al. 1993). From initial experiments performed in macaques and marmosets (HICKS et al. 1977; ALBRECHT et al. 1972,1980), cell-mediated immunity may be considered important in the control of MV infection in monkeys. In these experiments, treatment of rhesus monkeys with the immunosuppressive agent cyclophosphamide or with T lymphocyte depleting drugs resulted in enhanced and prolonged viremia of infected animals.

Recently, we have undertaken a series of experiments in cynomolgus monkeys to study the immune response parameters induced by different measles vaccines and vaccine candidates (VAN BINNENDIJK et al. 1994). In preliminary experiments, MV-seronegative cynomolgus monkeys were inoculated intratracheally with the wild-type strain BIL of MV, which was recently isolated and biologically cloned from a patient with acute measles in The Netherlands. All infected monkeys developed strong virus-specific IgM and long-lasting (virus-neutralizing) IgG serum antibodies (VAN BINNENDIJK et al. 1994). Furthermore, the long-term presence of MV-specific $CD4^+$ and $CD8^+$ T cells were detected in the PBMCs of most infected monkeys using techniques similar to those previously described for MV infections in humans (VAN BINNENDIJK et al. 1990) and for simian immunodeficiency virus (SIV) infections in rhesus monkeys (MILLER et al. 1990; YAMAMOTO et al. 1990). Cynomolgus monkey PBMCs stimulated in vitro with MV-infected B-LCLs resulted in the predominant expansion of primarily MV-specific $CD8^+$ T cells which were capable of destroying virus-infected target cells in vitro (VAN BINNENDIJK et al. 1994). Thus, these studies support the hypothesis that the induction of a class I MHC-restricted CTL response and memory by $CD8^+$ T cells, as a direct result of virus infection in vivo, may control MV infections by eliminating virus-infected cells in both humans and monkeys (VAN BINNENDIJK et al. 1990, 1992, 1994).

3.5 Measles Vaccination in the Presence of Maternal Antibody

Almost immediately after monkeys have been captured they are usually vaccinated with live attenuated measles vaccines in order to prevent the high incidence of measles during captivity due to human contacts. Usually good levels of protection will be achieved by vaccination and in most cases without the development of any clinical signs (WELSHMAN 1989). However, it may occur that some animals are already naturally infected with measles prior to vaccination (HALL et al. 1971). Previous studies on the efficacy of measles vaccination in a breeding colony of rhesus monkeys showed that captive-born offspring from wild-born MV-seropostive mothers were unable to respond to vaccination up to 6 months of age. Revaccination of this group from the age of 8-10 months onward resulted in significant MV-specific serum antibody titer rises (MACARTHUR et al. 1982). As in humans, MV-specific serum antibodies, representing maternal antibodies, were detected in the prevaccination sera of these animals and were detectable up to 9-12 months of age (MACARTHUR et al. 1979). In unvaccinated, newborn cynomolgus monkeys, maternal MV-specific serum antibody titers decreased linearly with a half-life time of approximately three and a half weeks. Furthermore it has been shown that these titers never exceeded the MV-specific antibody titers of the mothers (FUJIMOTO et al. 1983b). It is most likely that, as in human babies, the placental transfer of IgG is responsible for the MV-specific antibody titers in the newborns rather than the transfer of Immunoglobulin by breast-feeding (ALBRECHT et al. 1977b; FUJIMOTO et al. 1983a,b).

In developing countries measles is still responsible for high mortality rates among children during their first year of life. Indeed, it is this group of young children who in most cases cannot be vaccinated with live attenuated vaccines due to the interference of maternal antibodies with the attenuated virus replication. Therefore, a suitable monkey model in, e.g., cynomolgus or rhesus macaques would provide the opportunity to study the efficacy and safety of novel generations of measles vaccines able to induce a protective immune response in the presence of maternal antibodies.

4 Concluding Remarks

Several primate species can be infected experimentally with MV. Also spontaneous infections may occur after contacts with humans, after which MV may spread from monkey to monkey and cause serious losses in affected colonies or during the shipment of animals. In certain species, e.g., cynomolgus and rhesus macaques, infection with wild-type MV causes a disease of varying severity, quite similar to measles in humans. The development of antibody and cell-mediated immunity to MV largely parallels the development of MV-specific immunity in humans. Attenuated MV strains used in human vaccines also proved to be attenuated for primates and upon infection rendered the animals immune to subsequent wild-type MV infection. As in humans, MV-specific passively acquired maternal antibodies interfere with wild-type and attenuated MV infection in macaques. Unlike macaques, marmosets develop a usually fatal infection with wild-type MV upon experimental or contact infection. The infection is not characterized by the signs and symptoms of measles that are typical for the disease in humans. In contrast, experimental infection with attenuated MV vaccine strains does not lead to disease symptoms in marmosets.

Taken together, cynomolgus and rhesus macaques may be considered suitable animals for studying the pathogenesis of and immune-mediated protection from MV infection, induced by natural infection or vaccination. Especially in the light of the eradication strategy of measles, as recently adopted by the WHO, the present availability of a suitable macaque model to study the potential of novel generations of vaccines that may overcome the specific problems related to maternally derived antibodies and atypical measles syndrome is of major importance.

Acknowledgements. The authors wish to thank Conny Kruyssen and Miek Eskens for help in preparing this manuscript.

References

Albrecht P. Shabo AL Burns R, Tauraso NM (1972) Experimental measles encephalitis in normal and cyclophisphamide-treated Rhesus monkeys. J Infect Dis 126: 154–161

Albrecht P. Burnstein T, Klutch MJ, Hicks JT, Ennis FA (1977a) Subacute sclerosing panencephalitis: experimental infection in primates. Science 195: 64–66

Albrecht P, Ennis FA, Satzman EJ, Krugman S (1977b) Persistence of maternal antibody in infants beyond 12 months:mechanism of measles vaccine failure. J Pediatr 91: 715–722

Albrecht P, Lorenz D, Klutch MJ, Vicjers JH, Ennis FA (1980) Fatal measles infection in marmosets pathogenesis and prophylaxis. Infect Immun 27: 969–978

Albrecht P, Lorenz D, Klutch MJ (1981) Encephalitogenicity of measles virus in marmosets Infect Immun 34: 581–587

Bankamp B, Brinchmann UG, Reich A, Niewiesk S, ter Meulen V, Liebert UG (1991) Measles virus nucleocapsid protein protects rats from encephalitis J Virol 65: 1695–1700

Blake FG, Trask JD (1921) Studies on measles II. Symptomology and pathology in monkeys experimentally infected. J Exp Med 33: 413–432

Brinckmann UG, Bankamp B, Reich A, ter Meulen V, Liebert AG (1991) Efficacy of individual measles virus structural proteins in the protection of rats from measles encephalitis. J Gen Virol 72: 2491–2500

Burnet FM (1968) Measles as an index of immunological function. Lancet 2: 610–613

Burnstein T, Jensen JH, Waksman BH (1964) The development of a neurotropic strain of measles virus in hamsters and mice. J Infect Dis 114: 265–272

Carrigan DR (1986) Round cell variant of measles virus: neurovirulence and pathogenesis of acute encephalitis in newborn hamsters. Virology 148: 349–359

Carrigan DR, Kabacoff CM (1987) Nonproductive, cell-associated virus exists before the appearance of antiviral antibodies in experimental measles encephalitis Virology 156: 185–188

Chan SPK (1985) Induction of chronic measles encephalitis in C57B1/6 mice. J Gen Virol 66: 2071–2076
Contreras G, Furesz J (1992) Possible influence of measles virus infection of cynomolgus monkeys on the outcome of the neurovirulence test for oral poliovirus vaccines. Biologicals 20: 27–33
De Vries P, van Binnendijk RS, van der Marel P, van Wezel AL, Voorma HO, Sundquist B, UytdeHaag FGCM, Osterhaus ADME(1988) Measles virus fusion protein presented in immune-stimulating complex (ISCOM) induces haemolysis-inhibiting and fusion-inhibiting antibodies, virus-specific T cells and protection in mice. J Gen Virol 69: 549–559
Degwitz R (1927) Etiology of measles J Infect Dis 41: 304
Drillien R, Spehner D, Kirn A, Giraudon P, Buckland R, Wild F, Lecocq JP (1988) Protection of mice from fatal measles encephalitis by vaccination with vaccinia virus recombinants encoding either the hemagglutinin or the fusion protein. Proc Natl Acad Scl USA 85: 1252–1256
Enders JF (1962) Measles virus, historical review, Isolation and behaviour in various systems. Am J Dis Child 103: 282–287
Enders JF, Peebles TC (1954) Propagation in tissue cultures of cytopathogenic agents from patients with measles. Proc Soc Exp Biol Med 86: 277–286
Enders JF, Katz SL, Milovanovic MV, Holloway A (1960) Studies on an attenuated measles-virus vaccine. 1. Development and preparation of the vaccine: Technics for assay of effects of vaccination. N Engl J Med 263: 153–159
Fujimoto K, Terao K, Cho F, Honjo S (1983a) The placental transfer if IgG in the cynomolgus monkey. Japan J Med Sci Biol 36: 171–176
Fujimoto K, Terao K, Cho F, Honjo S (1983b) Immunoglobulins and measles antibody in sera of newborn cynomolgus monkeys and in milk of their mothers. Japan J Med Sci Biol 36: 209–214
Giraudon P, Wild TF (1985) Correlation between epitopes on hemagglutinin of measles virus and biological activities: passive protection by monoclonal antibodies is related to their hemagglutination inhibiting activity. Virology 144: 46–58
Goldberger J, Anderson JF (1911) An expermiental demonstration of the presence of the virus of measles in the mixed buccal and nasal secretions. J Am Med Assoc 57: 476–478
Good RA, Zak SJ (1956) Disturbances in gamma globulin synthesis as experiments of nature. Pediatrics 18: 109–149
Griffin DE, Mullinix J, Narayan O, Johnson RT (1974) Age dependence of viral experssion: comparative pathogenesis of two rodent-adapted strains of measles virus in mice. Infect Immun 9: 690–695
Habermann RT, Williams FP (1957) Diseases seen at necropsy of 708 Macaca multatta (rhesus monkey) and Macaca phillippinensis (cynomolgus monkey). Am J Vet Res 18: 419–426
Hall WC, Kovatch RM, Herman PH, Fox JG(1971) Pathology of measles in rhesus monkeys. Vet Pathol 8: 307–319
Herndon RM, Rena-Descalzi L, Griffin DE, Coyle PK (1975) Age dependence of viral expression. Electronic microscopic and immunoperoxidase studies of measles virus replication in mice. Lab Invest 33: 544–553
Hicks JT, Albrecht P (1976) Cytolytic, complement-dependent antibodies to measles virus in rhesus monkeys after administration of live or killed virus. J Infect Dis 133: 648–654
Hicks JT, Sullivan JL, Albrecht P (1977) Immune responses during measles infection in immunosuppressed rhesus monkeys. J Immunol 119: 1452–1456
Hime JM, Keymer IF (1975) Measles in recently imported colobus monkeys (Colobus quereza). Vet Rec 97: 392
Hurst EW, Cooke B (1941) Experimental measles: transmission of the disease to monkeys, failure to transmit measles to rabbits, cultivation of the virus on the chorio-allantoic membrane. Australian Medical Publishing Company, Sidney.
Inagawa DT, Adams JM (1958) Propagation of measles virus in suckling mice. Proc Soc Exp Biol Med 98: 567–569
Jacobson S, Richert JR, Biddison WE, Satinsky A, Hartzman RJ, McFarland HF (1984) Measles virus-specific T4 human cytotoxic T cell clones are restricted by class II HLA antigens. J Immunol 33: 754–757
Jacobson S, Rose JW, Flerlage ML, McFarlin DE, McFarland HF (1987) Induction of measles virus-specific human cytotoxic T cells by purified measles virus nucleocapsid and hemagglutinin polypeptides. Viral Immunol 1: 153–162
Janda Z, Norrby E, Marusyk H (1971) Neurotropism of measles virus variants in hamsters. J Infect Dis 124: 553–564
Josias A (1898) Recherches experimentales sur la transmissibilite de la rugeole aux animaux. Med Mod 9: 158
Katz SL (1965) Immunization of children with live attenuated measles vaccines: five years of experience. Arch Gesamte Virusforsch 16: 222–230

Katz SL, Enders JF (1959) Immunization of children with live attenuated measles vaccine. Am J Dis Child 85: 605–607
Katz SL, Milovanovic MV, Enders JF (1958) Propagation of measles virus in cultures of chick embryo cells. Proc Soc Exp Biol Med 97: 23–29
Kobune F, Skata H, Sugiura A (1990) Marmoset lymphoblastoid cells as a sensitive host for isolation of measles virus. J Virol 64: 700–705
Kraft W (1932) Zur Kritik der Experimentallen Masernübertragung. Z Gesamte Exp Med 81: 27
Kreth HW, ter Meulen V, Eckert G (1979) Demonstration of HLA restricted killer cells in patients with acute measles. Med Microbiol Immunol 165: 203–214
Levy BM, Mirkovic RR (1971) An epizootic of measles in a marmoset colony. Lab Anim Sci 21: 33–39
Liebert UG, ter Meulen V (1987) Virological aspects of measles virus induced encephalomyelitis in Lewis and BN rats. J Gen Virol 68: 1715–1722
Liebert UG, Linington C, ter Meulen V (1988) Induction of autoimmune reactions to myelin basic protein in measles encephalitis in Lewis rats. J Neuroimmunol 17: 103–118
Lorenz D, Albrecht P (1980) Susceptibility of tamarins (Saquinus) to measles virus. Lab Anim Sci 30: 661–665
Lucas CJ, Biddison WE, Nelson DL, Shaw S (1982) Killing of measles virus-infected cells by human cytotoxic T cells. Infect Immun 38: 226–232
Lucas WP, Prizer EL (1912) An experimental study of measles in monkeys. J Med Res 26: 181
MacArthur JA, Mann PG, Oreffo V, Scott GBD (1979) Measles in monkeys: an epidemological study. J Hyg 83: 207–212
MacArthur JA, Mann PG, Oreffo V, Scott GBD (1982) The effect of measles vaccination in rhesus monkeys (Macaca mulatta). Lab Anim 16: 267–270
Malvoisin E, Wild F (1990) Contribution of measles virus fusion protein in protective immunity: anti-F monoclonal antibody neutralize virus infectivity and protect mice against challenge. J Virol 64: 5160–5162
Mathiesen T, Hammarstrom L, Fridell E, Linde A, Wirsen G, Smith CI, Norrby E, Warhren B (1990) Aberrant IgG subclass distribution to measles in healthy seropositive individuals. In patients with SSPE and in immunoglobulin-deficient patients. Clin Exp Immunol 80: 202–205
Matumoto M, Saburi Y, Aoyama Y, Mutali M (1964) A neurotropic variant of measles in suckling mice. Arch Gesamte Virusforsch 14: 683–696
McChesney MB, Oldstone MB (1989) Virus-induced immunosuppression: infections with measles virus and human immunodeficiencey virus. Adv Immunol 45: 335–380
McChesney MB, Fujinama RS, Lerche NW, Marx PA, Oldstone MB (1989) Virus-induced immunosuppression: infection of peripheral blood mononuclear cells and suppression of immunoglobulin synthesis during natural measles virus infection in rhesus monkeys. J Infect Dis 159: 757–760
Meyer HM, Brooks BE, Douglas RD, Rogers NG (1962) Ecology of measles in monkeys. Am J Dis Child 103: 137–143
Miller MD, Lord CI, Stallard V, Mazzara GP, Letvin NL (1990).The gag-specific cytotoxic T lymphocytes in rhesus monkeys infected with the simian immunodeficiency virus of macaques. J Immunol 144: 122–128
Milovanovic MV, Enders JF, Mitus A (1957) Cultivation of measles virus in human amnion cells and developing chick embryo cells. Proc Soc Ex Biol Med 95: 120–127
Mitus A, Enders JF, Craig JM, Holloway A (1959) Persistence of measles virus and depression of antibody formation in patients with giant cell penumonia after measles. N Engl J Med 261: 882–889
Neighbour PA, Rager-Zisman B, Bloom BR (1978) Susceptibility of mice to acute and persistent measles infection. Infect Immun 21: 764–770
Niewiesk S, Brinckmann U, Bankamp B, Sirak S, Liebert UG, ter Meulen V (1993) Susceptibility to measles virus-induced encephalitis in mice correlates with impaired antigen presentation to cytotoxic T lymphocytes. J Virol 67: 75–81
Nil S, Kamahora J, Mori J, Takahashi M, Nishimura S Okuno Y (1964) Experimental pathology of measles in monkeys. Bikens J 6: 271–297
Norrby E (1985) measles vaccination, today and tomorrow. Ann Inst Pasteur/Virol 136E: 561–570
Norrby E, Oxman MN (1990) Measles virus. In: Fields BN (ed) Virology, vol I. Raven, New York, pp 1013–1044
Ohuchi R, Ohuchi M, Mifune K (1984) Slow development of measles virus (Edmonston strain) infection in the brain of nude mice. Microbiol Immunol 28: 757–764
Ohuchi R, Ohuchi M, Mifune K (1986) Protective effect of measles virus inoculation on subacute sclerosing panencephalitis virus-infected mice. Microbiol Immunol 30: 203–211
Ono K, Iwa N, Kato S, Konobe T (1970) Demonstration of viral antigen in giant cells formed in monkeys experimentally infected with measles virus. Bikens J 13: 329–337

Peebles TC, McCarthy K, Enders JF, Holloway A (1957) Behavior of monkeys after inoculation of virus derived from patients with measles and propagagted in tissue culture together with observations on spontaneous infections of these animals by agent exhibiting similar antigenic properties. J Immunol 78: 63–74
Plotz H (1938) La rougeole inapparente chez le single (*Macacus rhesus*) CR Soc Biol 133: 141
Potkay S, Ganaway JR, Rogers NG, Kinard R (1966) Epizootic of measles in a colony of Rhesus monkeys (*Macaca mulatta*). Am J Vet Res 27: 331–334
Remfry J (1976) A measles epizootic with 5 deaths in newly-imported rhesus monkeys (*Macaca mulatta*). Lab Anim 10: 49–57
Ruckle G, Rogers KD (1957) Studies with measles virus. II. Isolation of virus and immunologic studies in persons who have had natural disease. J Immunol 78: 341–355
Sakaguchi M, Yoshikawa Y, Yamamouchi K, Sata T, Nagashima K, Tekada K (1986) Growth of measles virus in epithelial and lymphoid tissues of cynomolgus monkeys. Microbiuol Immunol 30: 1067–1073
Sato TA, Kohama T, Sugiura A (1989) Protective role of human antibody to the fusion protein of measles virus. Microbiol Immunol 33: 601–607
Schaffer MF, Rake G, Stokes J Jr, O'Neil GC (1941) Studies on measles. II. Experimental disease in man and monkey. J Immunol 41: 241
Scott GBD, Keymer IF (1975) The pathology of measles in abyssinian colobus monkeys (*Colobus quereza*): a description of an outbreak. J Pathol 117: 229–233
Sergiev PG, Ryzantseva NE, Shroid IG (1960) The dynamics of pathological processes in experimental measles in monkeys. Acta Virol (Engl) 4: 265–273
Shishido A (1966) Natural infection of measles virus in laboratory monkeys. Jpn J Med Sci Biol 19: 221–222
Taniguchi T, Kamahora J, Kato S, Hagiwara K (1954) Pathology in monkeys experimentally infected with measles virus. Med J Osaka Univ 5: 367–396
UytdeHaag FGCM, van Binnendijk RS, Kenter MJH, Osterhaus ADME (1994) Cytotoxic T lymphocyte responses against measles virus. Curr Top Microbiol Immunol 189: 151–167
Van Binnendijk RS, Poelen MCM, de Vries P, Voorma HO, Osterhaus ADME, UytdeHaag FGCM (1989) Measles virus-specific human T cell clones. Characterization of specificity and function of $CD4^+$ helper/cytotoxic and $CD8^+$ cytotoxic T cell clones. J Immunol 142: 2847–2854
Van Binnendijk RS, Poelen MCM, Kuijpers KC, Osterhaus ADME, UytdeHaag FGCM (1990) The predominance of $CD8^+$ T cells after infection with measles virus suggests a role for $CD8^+$ class I MHC-restricted cytotoxic T lymphocytes (CTL) in recovery from measles. Clonal analyses of human $CD8^+$ class I MHC-restricted CTL. J Immunol 144: 2394–2399
Van Binnendijk RS, van Baalen CA, Poelen MCM, de Vries P Boes J, Cerundolo V, Osterhaus ADME, and UytdeHaag FGCM (1992) Measles virus transmembrane fusion protein synthesized de novo or presented in ISCOM is endogenously processed for HLA class I- and II-restricted cytotoxic T cell recignition. J Exp Med 176: 119–128
Van Binnendijk RS, van der Heijden RWJ, van Amerongen G, UytdeHaag FGCM, and Osterhaus ADME (1994) Viral replication and development of specific immunity after infection with different measles virus strains. J Inf Dis 170: 443–448
Waksman BH, Burnstein T, Adams RD (1962) Histologic study of the encephalomyelitis produced in hamsters by a neurotropic strain of measles. J Neuropathol Exp Neurol 21: 25–49
Welshman MD (1989) Measles in the cynomolgus monkey (*Macaca fascicularis*). Vet Rec 124: 184–186
Yamamoto H, Miller MD, Tsubota H, Watkins DI, Mazzara GP, Stallard V, Panicali DL, Aldovini A, Young RA, Letvin NL (1990) Studies of cloned simian immunodeficiency virus-specific T lymphocytes. J Immunol 144: 3385–3391
Yamanouchi K, Fukuda A, Kobune F, Hikita M, Shishido A (1969) Serologic survery with the sera of monkeys in regard to their natural infection with measles virus. Jpn J Med Sci Biol 22: 117–121
Yamanouchi K, Ageshira Y, Uchida N, Kodama H, Kobune F, Hayami M, Fukuda A, and Shishido A (1970) Giant cell formation in lymphoid tissues of monkeys inoculated with various strains of measles virus. Jpn J Med Sci Biol 23: 131–145
Yamanouchi K, Chino F, Kobune F, Kodama H, Tsuruhura T (1973) Growth of measles virus in the lymphoid tissues of monkeys. J Infect Dis 128: 795–799

Measles Virus Infections in Rodents

U.G. LIEBERT and D. FINKE

1 Introduction

Despite the availability of a safe and efficient live attenuated measles vaccine, which eliminated the fear of the highly contagious disease in industrialized countries, measles remains a severe disease in developing countries (see chapter by Clements and Cutts, this volume)

Moreover, our understanding of several important aspects of this infection have remained incomplete, particularly the pathogenesis of measles complications involving the CNS, the basis of measles virus-induced immunosuppression and the parameters of protective immune mechanisms.

As humans are the only natural host for measles virus, one of the major difficulties in studying measles has been the lack of a model for infection and disease. Monkeys, calves, lambs, dogs, ferrets, hamsters, mice and rats have been experimentally infected to study various pathogenetic aspects (FRAZER and MARTIN 1978; APPEL et al. 1981). Primates can be successfully infected with measles and have been used for many decades in infection experiments with clinical material (see chapter by van Binnendijk et al., this volume). However, the infection generally has a subclinical character and only marmosets and macaques develop severe disease and clinical symptoms similar to those in humans

Institut for Virology and Immunobiology, Universitty of Würzburg, Versbacher Strasse 7, 97078 Würzburg, Germany

(Albrecht et al. 1980). For these reasons macaques have primarily been used for immunological studies relevant to the development of novel measles vaccines. For pathogenetic studies, however, small laboratory animals, i.e., mice and rats, provide several advantages over the use of monkeys. Besides their easy and relatively inexpensive maintenance, their availability in sufficient quantities and in different inbred strains make rodents particularly suitable to study in detail the virus-host relationship, the parameters of antiviral immune response during experimental infection, the mechanism of immune protection, and the characteristics of immunopathologic events in measles complications. In this short review, virological and immunological findings of measles virus infections in rats and mice will be presented and their contribution to our understanding of pathogenetic aspects of measles CNS infections will be discussed.

2 Experimental Infection of Rodents

Different species and strains of rodents including mice, rats, hamsters and ferrets can be experimentally infected with measles virus. The replication of measles virus (MV), however, is largely restricted to neuroadapted strains and to the CNS of the animals (Imagawa and Adams 1958; Burnstein et al. 1964; Matumoto et al. 1964; Katz et al. 1968; Albrecht and Schumacher 1971; Janda et al. 1971; Griffin et al. 1974; Thormar et al. 1977; Chan 1985; Carrigan 1986; Ohuchi et al. 1984; Liebert and ter Meulen 1987). Neuroadaptation has been achieved by repeated intracerebral passage in newborn animals. Acute encephalitis is readily reproduced, but establishment of persistent infection in brain requires specific conditions (Katz et al. 1970). In ferrets and monkeys, these conditions include the use of cell-associated subacute sclerosing panencephalitis (SSPE) virus for intracerebral injection of animals with preinoculation passive immunity (Alberecht et al. 1977; Thormar et al. 1977; Ueda et al. 1975). Infection of newborn hamsters with virus from persistently infected cell cultures also produced a persistent infection in vivo (Norrby and Kristensson 1978), and in some cases persistent infection of hamsters led to the development of hydrocephalus, most likely due to infection of the ependymal cell layer (Haspel and Rapp 1975; Norrby et al. 1980).

Both infectious and defective cell-associated virus have been used successfully. Most commonly used MV strains for induction of acute encephalitis of mice and rats are either derived from isolates of SSPE patients (Byington et al. 1970; Wild et al. 1979), isolates from acute measles neuroadapted to hamster (Burnstein et al. 1964) or vaccine-derived virus neuroadapted to rat brain (Kobune et al. 1983). Intracerebral infection of susceptible inbred weanling rats and mice with the neurotropic CAM/RB measles virus leads to the development of acute and subacute (persistent) CNS disease processes (Liebert and ter Meulen 1987). Symptoms develop within 6–10 days after infection and consist of ruffled fur, hunched position, unsteady gait and ataxia, hindlimb paresis, convulsions which finally end in a lethargic state with inadequate feeding (inanition), severe weight loss and a moribund state. The acute MV-induced encephalitis is always fatal. Infectious virus can easily be recovered from brain but not from other organs including lungs, spleen, thymus, lymph nodes. In these organs it is also impossible to detect MV mRNA by northern analysis or in situ hybridization. Histopathologically, severe generalized encephalitis with destructive and infiltrative lesions in the entire brain parenchyma is found. For the characterization of viral determinants related to neurovirulence, measles virus variants were used. These were generated in vitro and in vivo under the selecting pressure of monoclonal antibodies. A region on the hemagglutinin (H) protein has been defined that may be important for the neurovirulence properties of CAM/RB (LIEBERT et al. 1994).

The outcome of the infection depends not only on the properties of the virus inoculated (JANDA et al. 1971; THORMAR et al. 1978; CARRIGAN 1986) but also on the route of infection. Intracerebral inoculation of Lewis rats or C57Bl/6 mice results in the most consistent reproduction of CNS disease, but intraperitoneal or intranasal infection may also lead to encephalitis in a minority of inoculated newborn to 2 day-old animals. Furthermore, susceptibility depends on the age at time of infection (GRIFFIN et al. 1974; HERNDON et al. 1975; LIEBERT and TER MEULEN 1987) and the strain of animals used (NEIGHBOUR et al. 1978; LIEBERT et al., unpublished). Generally, while newborn mice and rats uniformly succumb to acute encephalopathy[1] within a few days, older animals become increasingly less susceptible and develop either acute (AE) or subacute measles encephalitis (SAME). Among the parameters determining the susceptibility of rats and mice to experimental MV encephalitis are, in particular, the MHC genes. Lewis rats which carry the $RT1^l$ MHC haplotype are highly susceptible, and more than 75% of infected 3–4 week-old rats die from AE within 2 weeks following intracerebral inoculation. BN rats ($RT1^n$) are resistant with less than 15% developing acute disease. In the mouse model, strains carrying $H2^d$ or $H2^s$ (Balb/c, DBA2) are resistant while $H2^k$ or $H2^b$ (C3H, CBA or C57Bl/6 and C57Bl/10) are highly or moderately susceptible to infection with CAM/RB, respectively. The use of MHC congenic animals and F1 and F2 generations of susceptible and resistant animals confirmed that resistance to measles encephalitis in rats and mice is immunogenetically controlled.

In addition to AE, rats carrying RT1 haplotypes l or l x n may develop SAME with incubation periods of several weeks to months (LIEBERT and TER MEULEN 1987; LIEBERT et al. 1988). This disease occurs in up to 25% of 4-week-old CAM/RB-infected rats. Its course is monophasic and leads to death in about 40% of affected rats. The remainder undergo an apparent complete recovery from paresis, ataxia and occasional convulsions within 5–15 days. In these rats prominent inflammatory infiltrates composed of T and B lymphocytes and

[1] The term encephalopathy used to describe noninflammatory correlates of acute CNS disease with extensive destruction of the brain parenchyma and necrosis of all components of cortex and white matter as a consequence of the lytic virus-cell interaction seen in the absence of a functioning immune system. In acute or subacute measles encephalitis (SAME) inflammatory lesions prevail, and in SAME perivascular cuffing is the hallmark of the disease.

macrophages can be found, particularly around draining veins in subcortical and periventricular white matter. Primary demyelination does not occur, and destruction of brain parenchyma with neural cell loss is rather subtle. In resistant mouse or rat strains, instead of SAME a clinically silent encephalitis (CSE) may develop which can be diagnosed only by detection of characteristic histopathological lesions. Other factors that correlate with resistance are strength and specificity of the postinfection cell-mediated immunity to measles virus and the ability to synthesize and release interferon (IFN) following the experimental infection.

Other MV strains, particularly SSPE-derived or fresh wild-type isolates, display a very limited neurovirulence in Lewis rats. When, however, an SSPE-derived neurotropic strain of measles virus is used, Balb/c mice appear to be highly susceptible and die from intracerebral infection (WILD et al. 1990). Infection with the HNT virus induces in C57Bl/6 mice persistent infection and subacute disease in the majority of animals (CHAN 1985).

3 Virus-Host Relationship in the Brain of Mice and Rats

As described in detail elsewhere in this volume, the pathogenesis of MV infections of the human CNS is not fully understood. In SSPE and measles inclusion body encephalitis (MIBE) (see chapter by Schneider-Schaulies et al., this volume), a defective replication cycle is encountered in autopsy material, characterized by defective viral genomes that are unable to undergo a productive replication cycle even when transferred to permissive cells. However, the events which lead to the establishment of persistent infection, when and how the virus enters the brain, which receptor structures or intracellular factors determine brain cell tropism and susceptibility, and how the virus spreads in the brain in the absence of cell fusion and formation of virus particles are not known.

That both, host and viral factors are important for MV infection of the CNS has been shown by using the neurotropic HNT virus for infection of 3-week-old hamsters or of hamsters with a certain degree of passive immune protection by maternal antibodies (BYINGTON and JOHNSON 1972). In addition to AE, a persistent infection developed. Using monoclonal antibodies against different MV structural proteins, N and P proteins were always detected in brain cells during both acute and persistent infection, but M and H protein were expressed only during AE (JOHNSON et al. 1981). These findings were thought to reflect the ability of the host to restrict the infection while the immune system was unable to eliminate the virus from CNS tissue. Furthermore, the observations suggested the possibility of cell to cell spread in the infected CNS as a common mode of MV distribution to various regions of the brain.

In the rat model of intracerebral infection with CAM/RB, infectious virus could be recovered readily only during the initial 12–15 days postinfection, and no virus was reisolated after day 35 postinfection (LIEBERT and TER MEULEN 1987; LIEBERT et al., submitted). Usually, successful isolation required cocultivation of brain cells with noninfected permissive Vero cells. The titer of virus remained moderate and reached a maximum of 5×10^5 to 20×10^5 PFU/g brain weight. From animals with CSE to SAME no infectious virus could be recovered. Furthermore, already during the acute phase of infection and disease the expression of the envelope proteins M, F and H, in infected brain cells was reduced. Less than 20% of the infected cells expressed all major structural proteins (H, F, N, P, M) simultaneously, and in SAME this reduction was more pronounced. The virological data correlate with histopathological findings. In the brain of mice and rats giant cell formation is not detectable and no extracellular virus particles are found (VAN POTTELSBERGHE et al. 1979). In contrast, multinucleated giant cell formation is a hallmark of experimental measles infection in hamsters (RAINE et al. 1975). These findings suggest that the factors interfering with viral replication in the rat brain are similar in SSPE and in experimental rodent MV infection.

The molecular biological analysis of MV transcription in brain material from Lewis rats with AE and SAME revealed restricted expression of viral envelope proteins which was linked to an attenuated expression of the corresponding mRNAs (LIEBERT and TER MEULEN 1987; SCHNEIDER-SCHAULIES et al. 1989). In SAME, a specific translational defect was found for the M specific mRNA that was not related to sequence mutations (BACZKO et al. 1993). These transcriptional alterations were also observed in primary rat astroglial cells within hours following infection but not in other cell types, suggesting host cell-dependent control mechanisms of viral transcription (SCHNEIDER-SCHAULIES et al. 1990). In addition, the differentiation state of brain cells influences MV transcription, as seen in brain material of experimentally infected animals (LIEBERT et al. 1990b) and in tissue culture (YOSHIKAWA and YAMANOUCHI 1984). Attenuated MV transcription has also been found in the presence of virus-neutralizing antibodies in tissue culture experiments and in the animal model (FUJINAMI and OLDSTONE 1980; BARRETT et al. 1985; RAGER-ZISMAN et al. 1984; LIEBERT et al. 1990b; SCHNEIDER-SCHAULIES et al. 1992).

The in vitro observations largely explain the modifying effect on the disease course of passively transferred MV-neutralizing anti-H antibodies into infected newborn rats and mice (LIEBERT et al. 1990b; RAMMOHAN et al. 1983). Subacute diseases with prolonged incubation periods develop in these animals instead of the acute necrotizing encephalopathy. The molecular biological correlate is highly attenuated viral transcription in individual infected cells accompanied by a very low frequency of envelope-specific mRNAs and envelope protein expression. A remarkably similar pattern was detected in brain material from infected BN rats which naturally mount high titers of virus-neutralizing antibodies. However, in a hamster model it was shown that the presence of antibodies is not an absolute requirement for the establishment of persistent infection, since in the animals a spontaneous conversion from a lytic to a persistent infection was observed (CARRIGAN 1986; CARRIGAN and KABACOFF 1987). Furthermore, mutations in the nucleotide sequence or cap modifications of MV RNA, particularly the M gene-specific mRNA, have not been detected in rat CNS and hence are not required for MV persistence in brain tissue (Baczko et al., unpublished) despite the similarity of the defective viral replication cycle in the murine brain and in SSPE. This observation suggests that mutations of MV genes are apparently not essential for the initial establishment of persistent infection and rather develop thereafter. Therefore, cellular control mechanisms are likely to play an important role, as was indicated by inefficient MV protein synthesis in human glial cells after in vitro differentiation (SCHNEIDER-SCHAULIES et al. 1993a).

The infection of brain cells with MV also has important consequences for the expression of surface molecules on these cells, including adhesion molecules and MHC antigens (MASSA et al. 1987; OLSSON et al. 1987; GOGATE et al. 1991; KRAUS et al. 1992). MV infection was shown to lead to an increased expression of both class I and class II antigens and adhesion molecules such as ICAM-1 on cultivated glial cells. This process was IFN-γ-dependent and could be amplified by tumor necrosis factor-α (TNF-α). In vivo, where under normal conditions MHC antigens occur only in a low concentration in the CNS, a focal measles infection in rat brains resulted in up-regulation of these antigens throughout the entire brain (OLSSON et al. 1987). Expression of class I and II antigens not only occurred at the site of virus replication but also elsewhere in distant areas. Since antibodies against IFN-γ could not block class II induction, other as yet undefined cytokines or mechanisms of activation may play a role in vivo.

4 Parameters of the Antiviral Immune Response

In the rodent model, special consideration was given to immune parameters during experimental measles infection. In the absence of a complete immune system, a slowly developing brain infection was demonstrated in nude mice (OHUCHI et al. 1984). Reconstitution of T cells, but not of IFN or antiviral antibodies, completely prevented development of the disease (OHUCHI et al. 1986).

In acute MV-induced encephalitis in young rats and mice, which is always fatal, the effects of humoral and cellular immune responses have been analyzed in detail. Passive immunization of animals was shown to convert an acute infection into one of a persistent nature (GIRAUDON and WILD 1985; RAMMOHAN et al. 1983; LIEBERT et al. 1990b). The active induction (e.g., immunization with Iscom preparations containing the H and F proteins) or passive transfer of sufficient quantities of neutralizing antibodies completely suppressed the development of disease in newborn and weanling animals (VARSANYI et al. 1987; DE VRIES et al. 1988; MALVOISIN and WILD 1990; BRINCKMANN et al. 1991). Furthermore, the presence of maternal antibodies was shown to protect newborn animals from disease. These findings match the resistance to encephalitis observed in BN rats that mount an early high level of MV-specific humoral immune response (LIEBERT and TER MEULEN 1987). However, the presence of disease modifying antibody was shown to be generally insufficient to eliminate MV from the infected brain. The apparent explanation is that in the course of the experimental infection in rodents, viral envelope gene expression is restricted and virus particles are not released from infected brain cells. On the basis of these molecular biological characteristics it was anticipated that antibodies are not the prime factor in overcoming the CNS infection. Hence it was not surprising to find that even under conditions which virus-neutralizing antibodies are not induced, animals can be protected. This was demonstrated by immunization with recombinant vaccinia viruses expressing the internal nucleocapsid (N) protein, which prevented the disease that results from subsequent challenge with MV in nonimmunized rats (BANKAMP et al. 1991).

The role of the cell-mediated immune response, its effector components and its specificity were defined using recombinant vaccinia virus technology to express individual MV structural proteins. Following immunization of rats and mice with such recombinant vaccinia viruses, MV-specific antibodies and T cells were demonstrated. Challenge experiments with CAM/RB revealed a protective effect for all investigated susceptible strains of mice and rats of those recombinants that express either the MV H and F glycoproteins or the N protein (DRILLIEN et al. 1988; BANKAMP et al. 1991; BRINCKMANN et al. 1991; Finke et al., unpublished). MV was only isolatable for 5–7 days following infection with CAM/RB and histopathological changes were scarce. The kinetics of the humoral immune response supported the concept that antibodies are probably not involved in the elimination of MV from the brain because they either are unable to neutralize MV, as in the case of anti-N antibodies, or they become demonstrable only after the elimination of MV from the brain. Similar results were obtained in nude mice (OHUCHI et al. 1986) and in protection experiments using Iscom preparations containing H or F protein (VARSANYI et al. 1987).

The cellular antiviral immune response was analyzed further by two approaches. First, lymphocyte subpopulations were depleted by in vivo administration of monoclonal antibodies directed against the CD4 or CD8 surface molecules of lymphocytes (BANKAMP et al. 1991; FINKE and LIEBERT 1994). Second, MV-specific T cell lines were transferred adoptively into MV-infected rats. Both approaches rendered concordant results. Following intracerebral infection with CAM/RB those animals that had disposable CD4 T cells specific for N, H or F proteins were protected from the development of encephalitis. Moreover, when CD8 T cells were depleted, rats were still completely protected by adoptive transfer of CD4 T cells recognizing both internal N and surface envelope H and F viral proteins (REICH et al. 1992; LIEBERT et al. 1993). This treatment did not result in the timely local production of neutralizing antibody in the CNS. From the results it was concluded that both CD8 T cells and antibodies were not required for immune protection. These results were surprising because the susceptibility of mice to measles encephalitis was shown to correlate with their ability to generate a MHC class I restricted CD8 T cell-mediated cytotoxic immune response in vitro (NIEWIESK et al. 1993). However, in vivo this cell type is apparently dispensible for efficient elimination of MV and protection from disease in the murine encephalitis model. In rodent models a number of T and B cell epitopes have been defined (GIRAUDON et al. 1992; PARTIDOS and STEWARD 1990). It is, however, at present unknown where on the MV genome protective epitopes are located.

5 Mechanism of Immune Protection

The mechanism of the antiviral activity of CD4 T cells in vivo is still unknown. By using a genetic marker, it was shown in adoptive transfer experiments into MV infected rats that MV-specific CD4 T cells from a donor animal enter the brain of the host. These cells accumulated in infected areas where, however, in an immunocompetent animal they comprised only about 5% of all infiltrating T cells (KOLOKYTHAS and LIEBERT, unpublished). One mode of action of the virus-specific CD4 T cells would be a direct cytolysis of infected target cells, as has been demonstrated for human MV-specific peripheral blood lymphocytes in vitro (JACOBSON et al. 1986, 1989; VAN BINNENDIJK et al. 1989; LUCAS et al. 1982). Under special conditions MV-specific CD4 T cells from mice, and to a much lesser degree from rats, can also exhibit in vitro cytotoxic activity. However, it seems unlikely that MHC class II-restricted cytolysis plays a major role in eliminating MV from the brain, because the major cell population in the infected rodent brain consists of neurones which under normal conditions and during MV infection do not express MHC class II molecules. Furthermore, in the lymphocytic choriomeningitis virus (LCMV) infection model it was shown that virus-specific MHC class II-restricted cytolytic lymphocytes induce an immunopathological focus in the brain (MULLER et al. 1992). Another possibility of antiviral activity of CD4 T cells would be via local cytokine release and probably the activation of mononuclear phagocytes. Based on the observed paucity of virus-specific CD4 T cells in the recipient infected animal, it was concluded that recruitment of additional cells that exert antiviral effector functions are likely to constitute the essential factor in the combat of virus in an infected brain. Experiments using irradiated rats or treatment with immunosuppressive drugs, e.g. cyclosporine A, imply an important role for mononuclear phagocytes (Liebert et al. unpublished). The mechanism of recruitment might be explained by the ability of activated CD4 T cells to secrete cytokines. For example, IFN-γ and TNF enhance the expression of VCAM-1 on brain endothelium cells, to which stimulated T cells bind before they enter the brain and encounter viral antigen; this leads to further activation and subsequent secretion of cytokines (BARON et al. 1993). Cytokines delivered locally by CD4 T cells can contribute to selective virus elimination by either interacting directly with infected brain cells or by activating nonspecific effector cells including phagocytes (KARUPIAH et al. 1991; LUCCHIARI et al. 1993).

IFN-γ increases the expression of MHC class II molecules on astrocytes which can act as antigen presenting cells in the neighborhood of infected neurons (FONTANA et al. 1984; WONG et al. 1985; MASSA et al. 1987). In addition, IFN-γ is an important factor for the activation of macrophages and NK cells which can become potent IFN-γ producers on their own (TRINCHIERI and PERUSSIA 1985; PERUSSIA 1991). Activated NK cells can function as antiviral effector cells at an early time point of infection, when a virus specific T lymphocyte response still is absent. Furthermore, in the presence of IFN-γ preferential expansion of Th1 cells was observed in vitro (GAJEWSKI et al. 1989). It is conceivable, that the source of IFN-γ and TNF-α in MV infection of the murine CNS is CD4 T cells—at least under protective conditions. A detailed characterization in vitro revealed that all protective T cells lines produce high amounts of interleukin-2 (IL-2), IFN-γ and TNF-α but not IL-4 of IL-6 (Finke and Liebert, unpublished). This cytokine pattern is characteristic for CD4 T cells belonging to the Th1 helper subtype. In this context it is interesting that the predominant generation of Th2 cells, which is seen after both natural MV infection and vaccination against measles in humans, has led to the hypothesis that the lack of virus specific Th1 cells may contribute to the immunosuppression seen after infection or vaccination (GRIFFIN and WARD 1993; WARD and GRIFFIN 1993). Data obtained in the mouse model are consistent with this concept, as from the highly susceptible C3H mouse strain preferentially Th2 CD4 T cells, but not Th1 cells, can be isolated.

The importance of IFN-γ in the combat of MV CNS infection was also demonstrated in vivo. Ablation of IFN by treatment of infected mice with anti-IFN antibodies completely suppressed the protective effect of a functioning cell-mediated immune system and converted the resistent state of Balb/c mice into that of high susceptibility (Finke and Liebert, unpublished). Other cytokines are also involved, as was shown by the profound effect on the disease course and on mortality of treatment of animals which were resistant to MV infection, with anti-TNF-α antibodies.

Furthermore, it could be shown that the infection of brain cells leads to a differential induction of cytokines in primary and persistent MV infection in human glial cell cultures (SCHNEIDER-SCHAULIES et al. 1993b). In conclusion, cytokines secreted by Th1 cells may constitute an important parameter in elimination of MV from murine brain, while cytokines secreted by Th2 cells seem to suppress the cell-mediated immune response in clearance of infected tissue.

6 Autoimmunity and Immunopathology in the Infected Brain

Of particular pathogenetic interest is postinfectious encephalomyelitis (acute measles encephalomyelitis), a well known complication of many cases of acute measles (see chapter by Katz, this volume). In patients with this disease a significant proliferative response of isolated peripheral lymphocytes against myelin basic protein (MBP) was found during the clinical symptoms and for several weeks thereafter (JOHNSON et al. 1984). MV RNA and protein are absent from CNS and neuropathological abnormalities consist of perivenular demyelination. These observations suggest an immunopathological (autoimmune) basis for the development of measles encephalitis (JOHNSON and GRIFFIN 1986).

The most widely used model for autoimmune diseases of the CNS is experimental allergic encephalomyelitis (EAE) which can be induced in laboratory animals by injection of MBP or proteolipid (PLP) apoprotein in combination with adjuvants (RAINE 1984; YAMAMURA et al. 1986). EAE in rats is characterized by transient clinical signs including weight loss and weakness or paralysis of hind limbs and neuropathological changes consisting of inflammatory perivascular lymphomonocytic cuffing found predominantly in the white matter of the spinal cord. Susceptibility depends on the age of animals and the inoculum used (ITOYAMA and WEBSTER 1982; VANDENBARK et al. 1985; LEVINE and SOWINSKI 1973, 1975). EAE can be passively transferred from a diseased to a naive recipient animal by intravenous injection of CD4 lymphocytes specific for MBP. This treatment results in a CNS disorder of the host animal which is clinically and neuropathologically very similar to EAE (RAINE 1984). In analogy to EAE, the finding of a MBP-specific lymphoproliferative response was considered to be of pathogenetic importance in measles encephalitis. Below, the MV infection in a rat model with respect to autoimmune reactions and the mechanisms by which MV may alter host reactivity against self-antigens are discussed.

A role for MV as a cofactor in the development of EAE was suggested by early observations which showed that the course of EAE and its severity were potentiated in MV-infected hamsters (MASSANARI et al. 1979). In view of this observation it is interesting that two forms of SAME can be distinguished in Lewis rats. Type 1 SAME is characterized by persistent viral infection with focal gliosis and lymphomonocytic infiltrates restricted largely to infected brain areas. Here, viral antigen, MV mRNA and nucleocapsids were found but complete virion particles were absent, which is the reason for the failure to isolate infectious virus. In type 2 SAME the inflammatory process persists in the absence of MV antigen or viral nucleic acid. The lesions are very similar to those of naive rats receiving MBP-specific $CD4^+$ lymphocytes, and the infiltrates reveal a similar composition of lymphocyte subpopulations in SAME type 2 and EAE with a dominance of $CD4^+$ T cells over $CD8^+$ cells and a high proportion of macrophages (LIEBERT et al. 1988, 1990a; LIEBERT and TER MEULEN 1993). A humoral immune response to MBP was detected only in limited numbers of rats with SAME. In cerebrospinal fluid of the animals, in addition to measles-specific antibodies, oligoclonal bands of restricted heterogeneity were detected that probably react to brain antigens (DÖRRIES et al. 1988).

Based on these observations, the immune response to brain antigens was characterized in these animals. Splenic lymphocytes and superficial cervical lymph node cells were found to proliferate in vitro in the presence of MBP or PLP (LIEBERT et al. 1988). The intravenous transfer of MBP-reactive MHC class II-restricted $CD4^+$ T cell lines isolated from bulk cell populations induces a disease in naive syngeneic recipients with clinical and histopathological signs identical to T cell-mediated EAE. Analysis of the antigenic fine specificity revealed that MBP-specific T cell lines from both measles-infected and MBP challenged rats display an identical pattern of reactivity to a panel of synthetic peptides. These cells also responded to in vitro stimulation with peptide 69-84, which comprises the major encephalitogenic MBP sequence for Lewis rats (LIEBERT et al. 1990a). The high degree of antigenic specificity is further supported by the failure of the T cell lines to proliferate in the presence of disrupted measles virions, isolated MV proteins or other control antigens or peptide sequences. MV-specific T cell lines did not proliferate when MBP or synthetic MBP peptides were added to the cultures. Additionally, in some rats a positive reaction to the nonencephalitogenic peptide 69–81 was found in polyclonal spleen cell cultures from MV-infected rats. The role of this peptide in the induction of EAE-like disease was analyzed in experiments using a low dose of MV infection that failed to induce CNS diseases in animals. Histologically, there were no changes indicative of an active encephalitic disease process and viral antigen or RNA was not demonstrable in brain cells 4–8 weeks postinfection. When such animals were immunized with peptide 69–81, clinical and histological changes in the lumbar spinal cord developed which closely resembled EAE (LIEBERT et al. 1990a). Peptide 69–81 immunization of noninfected rats failed to induce disease or pathology. The distribution of the pathological changes was different from the MV-induced CNS lesions, which are mainly located in the gray and white matter of the cerebral hemispheres, midbrain and upper spinal cord. The disease induced by peptide 69–81 is not due to activation of MV in the brain of immunized Lewis rats, because virus could not be isolated from brain material and measles antigen was not detectable. The same result was obtained when peptide 69–81 specific T cells were transferred into MV-infected rats (LIEBERT and TER MEULEN 1993). The interaction between MBP peptide and MV infection was not observed when rats were infected intraperitoneally or when inactivated MV was used. The presence of MV-specific antibodies or T cells in peptide-immunized or peptide-T cell-transferred animals was not sufficient to result in the development of EAE, supporting the hypothesis that only after at least some initial active viral replication is the vulnerability of the brain to autoimmune aggression enhanced in the rat model.

If autoimmune mechanisms participate in the pathogenesis of virally induced encephalomyelitis, susceptibility to measles encephalitis and EAE should parallel in different rat strains depending on the genetic background. This is indeed the case, as BN rats, which are resistant to EAE and to measles encephalitis, did not develop a subacute clinical disease, although they were generally able to replicate MV (LIEBERT and TER MEULEN 1987). While the lymphoproliferative response to MBP is variable but represents a significant and reproducible phenomenon in Lewis rats, lymphocytes from BN rats were unable to proliferate in the presence of MBP (LIEBERT et al. 1990a). Obviously, susceptibility of rats to the development of MV-induced CNS changes and disease is multifactorial, with the development of a MBP-specific cell-mediated immune response representing a major factor.

There are several ways by which immune tolerance of self antigens during a viral infection can be inhibited:

1 Target cell change: During replication of viruses in living cells, host antigens are incorporated into the viral lipoprotein envelope. When progeny virus particles infect further cells these host antigens could be inserted into the cell membrane and exposed on the cell surface. Such antigens would be recognized as foreign by the host and would elicit an immune reaction in the same manner as any other previously unencountered protein (NOTKINS et al. 1984; SCHATTNER and RAGER-ZISMAN 1990). Furthermore, metabolic functions of host cells including enzymatic activities are altered during viral infections. Changes in peptidase activity in brain cells have been reported in experimental measles infection (Bever and Swoveland 1990). This is particularly intriguing for the development of autoimmune reactions in view of the hypothesis, that a hierarchy exists in the immunogenicity of various presented self-peptides: namely, dominant, sub-dominant, or cryptic, depending on their ability to prime for strong, weak, or no lymphoproliferation, as shown for hen egg lysozyme (Gammon et al. 1987; Cibotti et al. 1992). It is conceivable that changes in proteinase activity in virus-infected cells could result in processing of cryptic epitopes on autoantigens towards which the host would not be tolerant.

2 Interaction with lymphoid tissue: The interaction between lymphotropic viruses and components of the immune system in vivo could lead to imbalanced immune regulation and expansion of autoreactive lymphocyte clones. Epstein Barr-virus-transformed and immortalized human B lymphocytes may secrete, under certain conditions, autoantibodies that react to cellular constituents (Schattner and Rager-Zisman 1990). Infection of lymphocytes could form the basis of the cell-mediated autoimmune response in the rat model, since suppression of helper cell activity was observed in infected mouse T lymphocytes (McFarland 1974). In humans, it is well known that measles infection alters host immune functions including T cell responses of delayed-type hypersensitivity (DTH) and antibody production, and measles is sometimes followed by serious complications such as exacerbations of tuberculosis (Cherry 1987).

3 Molecular mimicry: Immunological cross-reaction between viral and normal host cell antigens may be another mechanism of autoimmunity (Oldstone and Notkins 1986). Computer analysis revealed that several viruses contain sequences identical or similar to MBP (Jahnke et al. 1985; Weise and Carnegie 1988). Moreover, the immunization of rabbits with a synthetic peptide from such a sequence of hepatitis B virus polymerase led to the induction of EAE lesions (Fujinami and Oldstone 1985). So far, encephalitogenic rat MBP sequences have not been found in the viral genome of MV, and measles-specific T cell lines isolated from infected rats do not proliferate in the presence of MBP (Liebert et al. 1988). However, one cannot exclude that this mechanism could play an important role in the development of autoimmune reactions if similar peptides of an infectious agent and a host cell protein share antigenic sites or interact with identical T cell receptors (Liebert and ter Meulen 1993).

4 Regulation of MHC antigens and autoimmunity: MHC antigens are usually expressed on most CNS parenchymal cell only at low levels or not at all (Wekerle et al. 1986; Sedgwick and Dörries 1991; Hart and Fabre 1981). Recent experiments, however, suggest that the CNS is effectively controlled by the immune system (Hickey and Kimura 1988; Hickey et al. 1991; Kolokythas and Liebert, submitted). Although the precise conditions are still unknown, the IFN dependent induction of MHC class II on brain cells in vivo (Traugott et al. 1985) and in vitro (Wong et al. 1985) and the subsequent pathological response such as a DTH reaction in genetically susceptible hosts is of particular interest. Astrocytes are effective antigen-presenting cells in vitro (Fontana et al. 1984), particularly for secondary $CD4^+$ T cell response (SEDGWICK et al. 1991; MATSUMOTO et al. 1986) and in inflammatory conditions in vivo microglial cells are the most abundant MHC class II expressing population (HAYES et al. 1987). With respect to MV, it has been shown that exposure of astrocytes in culture may lead to the expression of MHC class II antigen which can be further enhanced by the addition of TNF (MASSA et al. 1986, 1987). It is tempting to extrapolate these data to the in vivo situation. The problem, however, remains as to which factor(s) initiate(s) such a reaction in the CNS, since without the presence of MHC antigens, it is undoubtedly very difficult for T cells to recognize antigen, to become activated and to release lymphokines. Additionally, it is still unknown which role glial cells may play in the development of immune responses and there is no unequivocal evidence to suggest that antigen presentation to T cells in the brain actually occurs in vivo. Furthermore, the recent observation that in BN rats microglia are constitutively MHC class II positive suggests that mere MHC expression is not indicative of an increased susceptibility to inflammatory T cell responses in the CNS, as documented by the resistance to EAE and MV-associated encephalomyelitis (SEDGWICK et al. 1993). It is conceivable that glial MHC expression imparts on the cells the ability to interact with T cells and may down-regulate T cell responses rather than amplify them. This could be a determining factor for protection or development of an autoimmune disease process.

7 Concluding Remarks

The rodent models of experimental measles encephalitis add insight into and allow characterization of several aspects of the disease processes in the CNS associated with viral infections. The combined virological, molecular biological and immunological approach has revealed the nature of the interactions between virus and host tissue and between virus-specific and autoreactive immune reactions and the induction of immunopathology. The occurrence of cell-mediated autoimmune reactions to brain antigen as a result of MV infection is probably of pathogenetic importance. At present, no single factor can be pointed out which leads to this adverse response. Indeed, it is more conceivable that the autoimmune reaction is the sequelae of a number of different virus-induced changes, each one a relatively common event.

Acknowledgments. The experiments cited in this publication have been supported by the Deutsche Forschungsgemeinschaft, Bundesministerium für Forschung und Technologie and Hertie-Stiftung.

References

Albrecht P, Schumacher HP (1971) Neurotropic properties of measles virus in hamsters and mice. J infect Dis 124: 86–93

Albrecht P, Burnstein T, Klutch MJ, Hicks JT, Ennis FA (1977). Subacute sclerosing panencephalitis: experimental infection in primates. Science 195: 64–66

Albrecht P, Lorenz D, Klutch MJ, Vickers JH, Ennis FA (1980) Fatal measles infection in marmosets: pathogenesis and prophylaxis. Infect Immun 25: 969–978

Appel MJG, Gibbs EPJ, Martin SJ et al. (1981) Morbillivirus diseases of animals and man. In: Kurstak E (ed) Comparative diagnosis of viral diseases IV. Academic, New York, pp 239–297

Baczko K, Lampe J, Liebert UG, Brinckmann UG, ter Meulen V, Pardowitz I, Budka H, Cosby SL, Isserte S, Rima BK (1993) Clonal expansion of hypermutated measles virus in a SSPE brain. Virology 197: 188–195

Bankamp B, Brinckmann UG, Reich A, Niewiesk S, ter Meulen V, Liebert UG (1991) Protection from measles encephalitis of Lewis rats with vaccinia virus recombinants expressing the measles virus nucleocapsid protein. J Virol 65: 1695–1701

Baron JL, Madri JA, Ruddli NH, Hashim G, Janeway CA (1993) Surface expression of a4 integrin by CD4 T cells is required for their entry into brain parenchyma. J Exp Med 177: 57–68

Barrett PN, Koschel K, Carter M, ter Meulen V (1985) Effect of measles virus antibodies on a measles SSPE virus persistently infected C6 rat glioma cell line. J Gen Virol 66: 1411–1421

Bever CT, Swoveland P (1990) Astrocytosis and cathepsin D activity in experimental measles encephalomyelitis. Neurochem Res 15: 667–671

Brinckmann UG, Bankamp B, Reich A, ter Meulen V, Liebert UG (1991) Efficacy of single measles virus structural proteins in the protection of rats from measles encephalitis. J Gen Virol 72: 2491–2500

Burnstein T, Jensen JH, Waksman BH (1964) The development of a nurotropic strain of measles virus in hamsters and mice. J Infect Dis 114: 265–272

Byington DP, Johnson KP (1972) Expermiental subacute sclerosing panencephalitis in the hamster: correlation of age with chronic inclusion-cell encephalitis. J Infect Dis 126: 18–26

Byington DP, Castro AE, Burnstein T (1970) Adaptation to hamsters of neurotropic measles virus from subacute sclerosing panencephalitis. Nature 225: 554–555

Carrigan DR (1986) Round cell variant of measles virus: spontaneous conversion from productive to cell-associated state of infection. Virology 144: 337–350

Carrigan DR, Kabacoff CM (1987) Nonproductive, cell-associated virus exists before the appearance of antiviral antibodies in experimental measles encephalitis. Virology 156: 185–188

Chan SPK (1985) Induction of chronic measles encephalitis in C57Bl/6 mice. J Gen Virol 66: 2071–2076

Cherry JD (1987) Viral infections: measles. In: Feigin RD, Cherry JD (eds) Textbook of pediatric infectious diseases, vol 2. Saunders, Philadelphia. pp 1607–1635

Cibotti R, Kanellopoulos JM, Cabaniols JP, Halle-Panenko O, Kosmatopoulos K, Sercarz E (1992) Tolerance to a self protein involves its immunodominant but does not involve its subdominant determinants. Proc Natl Acad Sci USA 89: 416–420

De Vries P, van Binnendijk RS, van der Marel P, van Wezel AL, Voorma HO, Sundquist B, UytdeHaag FGCM, Osterhaus ADME (1988) Measles fusion protein presented in immune-stimulating complex (ISCOM) induced haemolysis-inhibiting and fusion-inhibiting antibodies virus-specific T cells and protection in mice. J Gen Virol 69: 549–559

Dörries R, Liebert UG, ter Meulen V (1988) Comparative analysis of virus-specific antibodies and immunoglobulins in serum and cerebro-spinal fluid of subacute measles virus-induced encephalomyelitis (SAME) in rats and subacute sclerosing panencephalitis (SSPE). J Neuroimmunol 19: 339–352

Drillien R, Spehner D, Kirn A, Giraudon P, Buckland R, Wild F, Lecocq JP (1988). Protection of mice from fatal measles encephalitis by vaccination with vaccinia virus recombinants encoding either the hemagglutinin or the fusion protein. Proc Natl Acad Sci USA 85: 1252–1256

Finke D, Liebert UG (1994) CD4 T cells are essential in overcoming experimental measles encephalitis. Immunology 83, in press

Fontana A, Fierz W, Wekerle H (1984) Astrocytes present myelin basic protein to encephalitogenic T cell lines. Nature 307: 273–276

Frazer KB, Martin SJ (1978) Measles virus and its biology. Academic, London.

Fujinami RS, Oldstone MBA (1980) Alterations in expression of measles virus polypeptides by antibody: molecular events in antibody-induced antigenic modulation, J Immunol 125: 78–85

Fujinami RS, Oldstone MBA (1985) Amino acid homolgy and immune responses between the encephalitogenic site of myelin basic protein and virus: a mechanism for autoimmunity. Science 230: 1043–1045

Gajewski TF, Schell SR, Nau G, Fitch FW (1989) Regulation of T cell activation: differences among T cell subsets. Immunol Rev. 111: 79–96

Gammon G, Shastri N, Cogswell J, Wilbur S, Sadegh-Nasseri S, Krzych W, Miller A, Sercarz EE (1987) The choice of T-cell epitopes utilized on a portein antigen depends on multiple factors distant from as well as at the determinant site. Immunol Rev. 98: 53–73

Giraudon P, Wild TF (1985) Correlation between epitopes on hemagglutinin of measles virus and

biological activities. Passive protection by monoclonal antibodies is related to their hemagglutination inhibiting activity. Virology 144: 46–58
Giraudon P, Buckland R, Wild TF (1992) The immune response to measles virus in mice. T-helper response to the nucleoprotein and mapping of T helper epitopes. Virus Res. 22: 41–54
Gogate N, Bakhiet M, Kristensson K, Norrby E, Olsson Y (1991) Gamma interferon expression and major histocompatibility complex induction during measles and vesicular stomatitis virus infections of the brain. J Neuroimmunol 31: 19–26
Griffin DE, Ward BJ (1993) Differential CD4 T cell activation in measles. J Infect Dis 168: 275–281
Griffin DE, Mullinix J, Narayan O, Johnson RT (1974) Age dependence of viral expression: comparative pathogenesis of two rodent-adapted strains of measles virus in mice. Infect Immun 9: 690–695
Hart DNJ, Fabre JW (1981) Demonstration and characterization of Ia-positive dendritic cells in the institial tissues of rat heart and other tissues, but not brain. J Exp Med 154: 347–361
Haspel MV Rapp F (1975) Measles virus: an unwanted variant causing hydrocephalus. Science 187: 450–451
Hayes GM, Woodroofe MN, Cuzner ML (1987) Microglia are the major cell type expressing MHC class II in human white matter. J Neurol Sci 80: 25–37
Herndon RM, Rena-Descalzi L, Griffin DE, Coyle PK (1975) Age dependence of viral expression. Electron microscopic and immunoperoxidase studies of measles virus replication in mice. Lab Invest 33: 544–553
Hickey WF, Kimura H (1988) Perivascular microglial cells of the CNS are bone marrow-derived and present antigen in vivo. Science 239: 290–292
Hickey WF, Hsu BL, Kimura H (1991) T-lymphocyte entry into the central nervous system. J Neurosci Res 28: 254–260
Imagawa DT, Adams JM (1958) Propagation of measles virus in suckling mice. Proc Soc Exp Biol Med 98: 567–569
Itoyama Y, Webster H de F (1982) Immunocytochemical study of myelin-associated glycoprotein (MAG) and basic protein (BP) in acute experimental allergic encephalomyelitis (EAE). J Neuroimmunol 3: 351–364
Jacobson S, Richert JR, Biddison WE, Salinsky A, Hartzman J, McFarland HF (1986) Measles virus-specific T4+human cytotoxic T cell clones are restricted by class II HLA antigens. J Immunol 133: 754–757
Jacobson S, Sehaly RP, Jacobson CL, McFarland HF, Long EO (1989) HLA class II-restricted presentation of cytoplasmatic measles virus antigen to cytotoxic T cells. J Virol 63: 1756–1762
Jahnke U, Fisher EH, Alvord EC (1985) Sequence homology between certain viral proteins and proteins related to encephalomyelitis and neuritis. Science 229: 282–284
Janda Z, Norrby E, Masuryk H (1971) Neurotropism of measles virus variants in hamsters. J Infect Dis 124: 553–564
Johnson KP, Norrby E, Swoveland P, Carrigan DR (1981) Experimental subacute sclerosing panencephalitis: selective disappearance of measles virus matrix protein from the central nervous system. J Infect Dis 144: 161–169
Johnson RT, Griffin D (1986) Virus-induced autoimmune demyelinating disease of the CNS. In: Notkins AL, Oldstone MBA (eds) Concepts in viral pathogenesis. Spinger, Berlin Heidelberg New York, pp 203–209
Johnson RT, Griffin DE, Hirsch RL, Wolinsky Js, Roedenbeck S, de Soriano IL, Vaisberg A (1984) Measles encephalomyelitis-clinical and immunologic studies. N Engl J Med 310: 137–141
Karupiah R, Woodhams GCE, Blanden RV, Ramshaw IA (1991) Immunbiology of infection with recombinant vaccinia virus encoding murine IL-2. Mechanisms of rapid viral clearance in immunocompetent mice. J Immunol 147: 4327–4332
Katz M, Rorke LB, Masland WS, Koprowski H, Tucker SH (1968) Transmission of encephalitogenic agents from brains of patients with subacute sclerosing panencephalitis to ferrets. N Engl J Med 279: 793–798
Katz M, Rorke LB, Masland WS, Barbanti-Brodano G, Koprowski H (1970) Subacute sclerosing panencephalitis: isolation of a virus encephalitogenic for ferrets. J Infect Dis 121: 188–195
Kobune K, Kobune F, Yamanouchi K, Nagashima K, Yoshikawa Y, Hayami M (1983) Neurovirulence of rat brain-adapted measles virus. Jpn J Exp Med 53: 177–180
Kraus E, Schneider-Schaulies S, Miyasaka M, Tamatani T, Sedgwick J (1992) Augmentation of major histocompatibility complex class I and ICAM-1 expression on glial cells following measles virus infection: evidence for the role of type-1 interferon. Eur J Immunol 22: 175–182
Levine S, Sowinski R (1973) Experimental allergic encephalomyelitis in inbred and outbred mice. J Immunol 110: 139–143

Levine S, Sowinski R (1975) Allergic encephalomyelitis in the reputedly resistant Brown Norway strain of rats. J Immunol 114: 597–601
Liebert UG, ter Meulen V (1987) Virological aspects of measles virus induced encephalomyelitis in Lewis and BN rats. J Gen Virol 68: 1715–1722
Liebert UG, ter Meulen V (1993) Synergistic interaction between measles virus infection and MBP peptide-specific T cells in the induction of EAE in Lewis rats. J Neuroimmunol 46: 217–224
Liebert UG, Linington C, ter Meulen V (1988) Induction of autoimmune reactions to myelin basic protein in measles virus encephalitis in Lewis rats. J Neuroimmunol 17: 103–118
Liebert UG, Hashim GA, ter Meulen V (1990a) Characterization of measles virus-induced cellular autoimmune reactions against myelin basic protein in Lewis rats. J Neuroimmunol 29: 139–147
Liebert UG, Schneider-Schaulies S, Baczko K, ter Meulen V (1990b) Antibody-induced restriction of viral gene expression in measles encephalitis in rats. J Virol 64: 706–713
Liebert UG, Reich A, Bankamp B, Brinckmann UG, ter Meulen V (1993) Control of measles virus infections by virus-specific CD4+ T cells. In: Thomas DB (ed) Viruses and cellular immune-responses. Dekker, New York, pp 279–291
Liebert UG, Flanagan S, Löffler S, Baczko K, ter Meulen V, Rima BK (1994) Antigenic determinants of measles virus haemagglutinin associated with neurovirulence. J Virol 68(3) (in press)
Lucas CJ, Biddison WE, Nelson DL, Shaw S (1982) Killing of mealses virus-infected cells by human cytotoxic T cells. Infect Immun 38: 226–232
Lucchiari MA, Modolell M, Eichmann K, Pereira CA (1993) In vivo depletion of interferon-gamma leads to susceptibility of A/J mice to mouse hepatitis virus 3 infection. Immunobiology 185: 457–482.
McFarland H (1974) The effect of measles virus infection on T and B lymphocytes in the mouse: suppression of helper cell activity. J Immunol 113: 1978–1983
Malvoisin E, Wild F (1990) Contribution of measles virus fusion protein in protective immunity: anti-F monoclonal antibody neutralize virus infectivity and protect mice against challenge. J Virol 64: 5160–5162
Massa PT, Dörries R, ter Meulen V (1986) Viral particles induce la antigen expression on astrocytes. Nature 320: 543–546
Massa PT, Schimpl A, Wecker E, ter Meulen V (1987) Tumor necrosis factor amplifies measles virus-mediated la induction on astrocytes. Proc Natl Acad Sci USA 84: 7242–7245
Massanari RM, Paterson PY, Lipton HL (1979) Potentiation of experimental allergic encephalomyelitis in hamsters with persistent encephalitis due to measles virus. J Infect Dis 139: 297–303
Matsumoto Y, Hara N, Tanaka R, Fujiwara M (1986) Immunohistochemical analysis of the rat central nervous system during experimental allergic encephalomyelitis, with special reference to la-positive cells with dendritic morphology. J Immunol 136: 3668–3676
Matumoto M, Saburi Y, Aoyama Y, Mutai M (1964) A neurotropic variant of measles in suckling mice. Arch Gesamte Virusforsch 14: 683–696
Muller D, Koller BH, Whitton JL, Lapan KE, Brigman KK, Frelinger JA (1992) LCMV-specific, class II restricted cytotoxic T cells in β_2-microglobulin-deficient mice. Science 255: 1576–1578
Neighbour PA, Rager-Zisman B, Bloom BR (1978) Susceptibility of mice to acute and persistent measles infection. Infect Immun 21: 764–770
Niewiesk S, Bankamp B, Brinckmann UG, Sirak S, ter Meulen V, Liebert UG (1993) Susceptibility to measles-induced encephalitis in mice correlates with imparied antigen presentation to CTL. J Virol 67: 75–81
Norrby E, Kristensson K (1978) Subacute encephalitis and hydrocephalus in hamsters caused by measles virus from persistently infected cell cultures. J Med Virol 2: 305–317
Norrby E, Swoveland P, Kristensson K, Johnson KP (1980) Further studies on subacute encephalitis and hydrocephalus in hamsters caused by mealses virus from persistently infected cell cutures. J Med Virol 5: 109–116
Notkins AL, Onodera T, Prabhaker BS (1984) Virus-induced autoimmunity. In: Notkins AL, Oldstone MBA (eds) Concepts in viral pathogenesis, Springer, Berlin, Heidelberg, New York, pp 210–215
Ohuchi R, Ohuchi M, Mifune K (1984) Slow development of measles virus (Edmonton strain) infection in the brain of nude mice. Microbiol Immunol 28: 757–764
Ohuchi M, Ohuchi R, Mifune K (1986) Protective effect of measles virus inoculation on subacute sclerosing panencephalitis virus-infected mice. Microbiol Immunol 30: 203–211
Oldstone MBA, Notkins AL (1986) Molecular mimicry. In: Notkins AL, Oldstone MBA (eds) Concepts in viral pathogenesis, Springer Berlin, Heidelberg, New York, pp 195–202
Olsson T, Maehlen J, Love A, Klareskog L, Norrby E, Kristensson K (1987) Induction of class I and class II transplantation antigens in rat brain during fatal and non-fatal measles virus infection. J Neuroimmunol 16: 215–224

Partidos CD, Steward MW (1990) Prediction and identification of a T cell epitope in the fusion protein of measles virus immuno-dominant in mice and humans. J Gen Virol 71: 2099–2105
Perussia B (1991) Lymphokine activated killer cells, NK cell, and cytokines. Curr Opin Immunol 3: 49–55
Rager-Zisman B, Egan JE, Kress Y, Bloom BR (1984) Isolation of clod-sensitive mutants of measles virus from persistently infected murine neuroblastoma cells. J Virol 51: 845–855
Raine CS (1984) Biology of disease. Analysis of autoimmune demyelination: its impact upon multiple sclerosis. Lab Invest 50: 608–635
Raine CS, Byington DP, Johnson KP (1975) Subacute sclerosing panencephalitis in the hamster: ultrastructure of the acute disease in newborns and weanlings. Lab Invest 33: 108–116
Rammohan KW, Dubois-Dalcq M, Rentier B, Paul J (1983) Experimental models to study measles virus persistence in the nervous system, Prog Neuropathol 5: 343–372
Reich A, Erlwein O, Niewiesk S, ter Meulen V Liebert UG (1992) Control of measles encephalitis in the rat by virus specific CD4⁺ T Cells. Immunology 76: 185–191
Schattner A, Rager-Zisman B (1990) Virus-induced autoimmunity. Rev Infect Dis 12: 204–222
Schneider-Schaulies S, Liebert UG, Baczko K, Cattaneo R, Billeter M, Meulen ter V (1989) Restriction of measles virus gene expression in acute and subacute encephalitis of Lewis rats. Virology 171: 525–534
Schneider-Schaulies S, Liebert UG, Baczko K, ter Meulen V (1990). Restricted expression of measles virus in primary rat astroglial cells. Virology 177: 802–806
Schneider-Schaulies S, Liebert UG, Segev Y, Rager-Zisman B, Wolfson M, ter Meulen V (1992) Antibody-dependent transcriptional regulation of measles virus in persistently infected neural cells. J Virol 66: 5534–5541
Schneider-Schaulies S, Schneider-Schaulies J, Bayer M, Loffler S, ter Meulen V (1993a) Spontaneous and differentiation-dependent regulation of measles virus expression in human glial cells. J Virol 67: 3375–3383
Schneider-Schaulies J, Schneider-Schaulies S, ter Meulen V (1993b) Differential induction of cytokines by primary and persistent MV-infection in human glial cells. Virology 195: 219–228
Sedgwick JD, Dorries R (1991). The immune system response to viral infection of the CNS. Semin Neurosci 3: 93–100
Sedgwick JD, Mößner R, Schwender S, ter Meulen V (1991) MHC-expressing non-hematopoietic astroglial cells prime only CD8+ T lymphocytes: astroglial cells as perpetuators but not initiators of CD4⁺ T cell responses in the central nervous system. J Exp Med 173: 1235–1246
Sedgwick JD, Schwender S, Gregersen R, ter Meulen V (1993) Resident macrophages (ramified microglia) of the adult brown Norway rat central nervous system are constitutively major histocompatibilitiy complex class II positive. J Exp Med 177: 1145–1152
Thormar H, Arnesen F, Mehta PD (1977) Encephalitis in ferrets caused by a nonproductive strain of measles virus (D.R.) isolated from a patient with subacute sclerosing panencephalitis. J Infect Dis 136: 229–238
Thormar H, Mehta PD, Brown HR (1978) Comparison of wild-type and subacute sclerosing panencephalitis strains of measles virus. J Exp Med 148: 674–691
Traugott U, Scheinberg LC, Raine CS (1985). On the presence of Ia-positive endothelial cells and astrocytes in multiple sclerosis lesions and its relevance to antigen presentation. J Neuroimmunol 8: 1–14
Trinchieri G, Perussia B (1985) Immune interferon: a pleiotropic lymphokine with multiple effects. Immunol Today 6: 131–136
Ueda S, Otsuka T, Okuno Y (1975) Experimental subacute sclerosing panencephalitis (SSPE) in a monkey by subcutaneous inoculation with a defective SSPE virus. Biken J 18: 179–181
Van Binnendijk RS, Poelen MCM, de Vries P, Voorma HO, Osterhaus ADME, Uytdehaag FGCM (1989) Measles virus-specific human T cell clones. J Immunol 142: 2847–2854
Van Pottelsberghe C, Rammohan KW, McFarland HF, Dubois-Dalcq M (1979) Selective neuronal, dendritic, and postsynaptic localization of viral antigen in measles-infected mice. Lab Invest 40: 99–108
Vandenbark AA, Gill T, Offner H (1985) A myelin basic protein-specific T cell line that mediates experimental allergic autoimmune encephalomyelitis. J Immunol 135: 223–228
Varsanyi TM, Morein B, Love A, Norrby E (1987) Protection against lethal measles virus infection in mice by immune-stimulating complexes containing the hemagglutinin or fusion protein. J Virol 61: 3896–3901
Ward BJ, Griffin DE (1993) Changes in cytokine production after measles virus vaccination: predominant production of IL-4 suggests induction of a Th_2 response. Clin Immunol Immunopathol 67: 171–177
Weise MJ, Carnegie PR (1988) An approach to searching protein sequences for superfamily

relationships or chance similarities relevant to the molecular mimicry hypothesis: application to the basis proteins of myelin. J Neurochem 51: 1267–1273

Wekerle H, Linington C, Lassmann H, Meyermann R (1986) Cellular immune reactivity within the CNS. Trends Neurosci 9: 271–275

Wild TF, Giraudon P, Bernard A, Huppert J (1979) Isolation and characterization of defective measles virus from a subacute sclerosing panencephalitis patient. J Med Virol 4: 103–114

Wild TF, Giraudon P, Spehner D, Drillien R, Lecocq JP (1990) Fowlpox virus recombinant encoding the measles virus fusion protein: protection of mice against fatal measles encephalitis. Vaccine 8: 441–442

Wong GHW, Bartlett PF, Clark-Lewis I, McKimm-Breschkin JL, Schrader JW (1985) Interferon-g induces the expression of H-2 and Ia antigens on brain cells. J Neuroimmunol 7: 255–278

Yamamura T, Namikawa T, Endoh M, Kunishita T, Tabira T (1986) Experimental allergic encephalomyelitis induced by proteolipid apoprotein in Lewis rats. J Neuroimmunol 12: 143–153

Yoshikawa Y, Yamanouchi K (1984) Effects of papaverine treatment on replication of measles virus in human neural and nonneural cells. J Virol 50: 489–496

The Paradigms of Measles Vaccinology

E. Norrby

1 Introduction

Measles is an eradicable disease. As in the case of smallpox, humans are the only natural host and the epidemic spread of both these diseases can only be maintained as a chain of serial direct transmissions of virus, involving acutely infected individuals. Furthermore the viruses causing the two diseases are monotypic and effective immunity can be induced by use of live vaccine viruses in each case. Whereas smallpox has been effectively eradicated for more than 15 years, measles continues to be a scourge in developing countries in spite of extensive vaccination campaigns. It is estimated that yearly one and a half million children die of the disease. This situation is unacceptable in the perspective of the availability of an effective vaccine, as has been highlighted by the World Health Organization (WHO) and other international organizations. More vigorous efforts towards regional and eventual global eradication of the disease and its causative agent are required. As a first goal a reduction of the number of measles cases by 95% and a reduction of disease-associated mortality from 1%-25% to 0,01%-0,1% has been targeted for 1995 by WHO. The question is if this can be reached by use of existing vaccines and by application of the current strategies. Available vaccines are highly efficacious and safe but since they represent replicating

MTC, Karolinska Institute, P.O. BOX 280, 171 77 Stockholm, Sweden immunogens they cannot be used at an early age, due to the presence of transplacentally acquired maternal antibodies. This fact has influenced the logistics of use of available vaccines. It has also prompted the question whether there is any approach that would allow the barrier of maternal antibodies, at least when being of a moderate level, to be overcome.

In industrialized countries the vaccine is not used before the age of 15 months. Under these conditions a categorical replication of vaccine virus is permitted. The take rate of vaccination under these conditions generally is about 90% and should optimally reach 95%. The immunity is durable but may wane in individuals with a low initial post-vaccination immunity. The immunity established after successful vaccination is not absolute in that subclinical replication of wild virus may occur after exposure. In some instances, secondary vaccine failures have also been identified. However, there is no evidence that virus replicating in a partially immune individual in a way to give no or mild symptoms can spread to cause disease in another unprotected individual (EDMONDSON et al.1990). Immunization schemes including two separate vaccinations are a means of safe-guarding an efficient herd immunity. The extreme contagiousness of the virus necessitates minimizing of pockets of nonimmunized groups of individuals in a population in order to prevent spread of virus.

In developing countries the situation is different from that in industrialized countries. The natural endemic occurrence of the infection causes an early exposure in life. In the context of other circumstantial influences a severe, occasionally fatal, disease may ensue. Thus, this situation calls for an early vaccination, ideally already at the age of 6 months or earlier. The window of susceptibility to measles in developing countries needs to be closed.

Since the take rate of existing live vaccines is progressively reduced in younger age groups, there is a limit to the earliest age at which vaccination can be given. Early vaccination prompts a follow up vaccination some 6–9 months later, since a fraction of the vaccines will represent primary vaccine failures. Such two-dose programs may not be readily pursued in developing countries. Regrettably the efficacy of vaccination may be further reduced in such countries due to problems with, e.g., stability of vaccines.

As one means of attempting immunization early in life, high-titer vaccine products, including certain strains of virus, were evaluated. It was found that increasing the dose of virus did in fact improve the take rate of vaccination. However, it was also discovered, unexpectedly, that immunized children had an increased mortality 6–12 months after vaccination. This mortality was due to intercurrent diseases and primarily afflicted girls. The cause of this tragic complication has not been defined but it is tempting to speculate that some kind of extended immune suppression caused by the replicating vaccine virus might have played a role. (see chapter by CLEMENTS and CUTTS).

The logistics of vaccine distribution obviously influence the eventual efficacy with reference to herd immunity. Potentially mass campaigns as employed, e.g., in Brazil may turn out to be highly efficacious in eliminating circulation of virus. Aspects within the realms of logistics of vaccinations, however, will not be considered in this chapter. Instead the possibilities for improvement upon available vaccines and for generation of qualitatively new kinds of products will be discussed. Some key, partly mythical and partly factual, problems in measles vaccinology will be intertwined in the discourse. The following questions reflect some aspects of these problems:

1. How categorical are the monotypic properties of the virus or, expressed differently, does strain variations (see chapter by RIMA et al.) have any significance in enhancing epidemic spread of virus in an immune population?

2. What qualitative and quantitative aspects of long-term immunity can be considered in evaluating epidemiologically significant potential contagiousness in individuals with varying degrees of immunity?

3. Which quantitative and qualitative phenomena guide the possibilities for an attenuated virus to overcome the barrier of maternal antibodies and is there a difference in the capacity of wild virus and of currently available vaccine virus to overcome this barrier?

2 The Marvel of the Life-Long Immunity After a Measles Virus Infection and the Enigma of Virus Monotypicity

Under normal conditions measles is a once in a life-time experience in unimmunized individuals. The solid long-time production of immunity is dependent on humoral and/or cell-mediated immune responses. Both arms of the immune defense can mediate long-term protection but cell-mediated phenomena have a dominating role in clearence of infections. There are three observations which highlight a major role for antibody in the prevention of infection. Firstly, maternally derived IgG1 and IgG4 protect newborn children against infection with measles virus. This immunity lasts until about 15 months of age but it is probably at an average shorter if the maternal immunoglobulin derives from vaccination-based immunity. Secondly, immunoglobulin-compromised children can be protected against measles when treated with a regular regime of immunoglobulin. Finally a child given immunoglobulin within 4–5 days after exposure to wild virus is capable of aborting the infection.

The targets for protective antibody responses to measles have been defined. Antibodies to the hemagglutinin (H) protein play a major role (VARSANYI et al. 1984, 1987). Hemagglutinating inhibiting (HI) monoclonal antibodies provide passive protection against encephalitis in rodents (GIRAUDON and WILD 1985). They may act by preventing virus attachment or by formation of complexes with virus particles, which when processed intercellularly do not lead to expression of virus infectivity. The H protein functionally interacts intimately with the second envelope glycoprotein, the fusion (F) component, during early stages of virus replication (WILD et al. 1991). In fact there are H-specific monoclonal antibodies that do not block hemagglutination, but still inhibit viral hemolysis and implicitly cell fusion (DE VRIES et al. 1987; Norrby et al. unpublished data). Antibodies against the F protein induced by hyperimmunization show no or only very weak in vitro neutralizing activity (Varsanyi et al. 1987). However if all HI antibodies in human convalescent sera are removed by absorption with H antigen a considerable fraction of the neutralizing activity remains (Norrby and Gollmar 1975). It has been inferred that F-specific antibodies have a significant protective role in vivo. Protection against a lethal infection in rodents was established by use of the F component associated with immunostimulating complexes (de Vries et al. 1988; Varsanyi et al. 1987) and of vaccinia virus vectors encoding the measles F gene (Drillien et al. 1988; Brinckmann et al. 1991). In dogs it was demonstrated that immunization with canine distemper virus F protein (Norrby et al. 1986) or nonreplicating whole measles virus (Appel et al. 1984) gave an immunity of infection-permissive character. Similar effects were seen in dogs immunized with vaccinia virus recombinants expressing the measles virus F protein (Taylor et al. 1991b, 1992). Thus it would seem that F-specific antibodies acting on their own generally appear to mediate infection-permissive immunity, but when they act together with the anti-H antibodies, they contribute significantly to a state of robust protection. High titers of H-specific antibodies have been interpreted to give complete protection on their own, but when they occur in lower concentrations they only give partial protection. In the latter situation they may, in the absence of functional anti-F antibodies, even be instrumental in development of atypical measles, an immune pathological complication seen after the use of formalin-inactivated measles vaccine. In order to guarantee the absence of such complications antibody responses to both surface glycoproteins are required (Norrby et al. 1975).

There is no evidence that antibody response to internal virion components play a critical role in antibody-mediated immunity. However, the importance of the various components is, expectedly, different when it comes to cell-mediated immune phenomena. Children with Bruton's sex-linked agammaglobulinemia can effectively manage a measles virus infection and acquire life-long immunity. The target for cell-mediated immunity has been dissected by use of vaccinia vectors carrying one or more different measles virus genes and by use of short peptides representing various sites in the viral proteins. The nucleocapsid (N) protein appears to play a prime role, but also the surface glycoproteins may have some involvement (Bankamp et al. 1991; Beauverger et al. 1993 and personal communication; Niewiesk et al. 1993; van Binnendijk et al. 1992). There seems to be a lesser involvement of other internal components such as the phosphoprotein (P) and the matrix (M) protein in cell-mediated reactions.

The term protection needs to be further defined. Clearly, a complete barrier immunity allowing no replication of an infecting virus might be established, e.g. during the immediate period after a wild virus infection. However, after some time the level of immunity has waned to allow a limited, probably mucosal, virus replication. This replication does not cause any symptoms and the infected individual does not appear to be a vehicle for transmission of virus. Similarly, there is no evidence for transmission of viruses from the rare cases of secondary vaccine failures associated with symptoms of varying intensity in immunized individuals. It can be speculated that the long-term contribution to immunity of specific cell-mediated functions, as compared to antibody-mediated protection, may be relatively weaker after vaccination than after a wild virus infection.

Measles virus, like any other single-stranded RNA virus, shows a high propensity for mutagenic changes. As already mentioned the virus categorically behaves as a monotypic agent. Minor antigenic differences have been observed in neutralization tests comparing laboratory strains and fresh isolates from cases of acute measles (ALBRECHT et al. 1981). Variations in occurrence of single epitopes in different proteins have been identified by use of monoclonal antibodies (CARTER et al. 1982; SHESHBERADARAN et al. 1983). Such variations have not only been encountered in the surface glycoprotein H but also in the M (SHESHBERADARAN et al. 1982) and in the N proteins (GIRAUDON et al. 1988; SATO et al. 1985). Using available reagents so far no epitope variations have been detected in the F component of various strains (SHESHBERADARAN et al. 1983). Propagation of virus in the presence of a neutralizing monoclonal antibody causes a rapid emergence of virus variants replicating in the presence of such an antibody (BIRRER et al. 1981). In fact, it is possible to select variants lacking many different epitopes with a potential to mediate neutralization (SHESHBERADARAN and NORRBY 1986). Such a variant remains neutralizable with polyclonal antiserum. Recently molecular epidemiological studies were performed in which nucleotide sequences of homologous regions of genes were compared. It was documented that strains with distinctive characteristics could circulate simultaneously during an epidemic (TAYLOR et al. 1991a) and that the viral genome shows a time correlated evolvement (ROTA et al. 1992). It should be emphasized that in spite of these observations measles virus strains, independent of their global geographic origin, react in a similar way with polyclonal antisera. Thus, apparently the massive selective pressure represented by postdisease immunity has not managed over many centuries to spurn a new type of measles virus. This fact must reflect some inherent functional constringency in properties of measles virus proteins and in this characteristic a truly monotypic virus like measles distinguishes itself from, e.g., orthomyxoviruses and lentiviruses.

3 The Three Avenues Towards New Measles Vaccines

Advances in the fields of molecular genetics, immunobiology, and immunochemistry regarding morbilliviruses have provided possibilities for development of new kinds of measles vaccines. In the process of development of such a new vaccine, several issues may be considered. However, it seems that among these issues there is a major emphasis on possibilities to generate a product for use in the young infant in the presence of maternal antibodies. Additional issues may also be raised. One such issue is stability, but this problem should be possible to solve by technical modifications not demanding alterations of the conceptual characteristics of the product. Another issue is the duration and efficacy of the vaccine-induced immunity. Admittedly, secondary vaccine failures have occurred, but these are rare events and they do not detract from the potential effective establishment of herd immunity. Claims of yielding herd immunity in generalized terms reflect deficient logistics in the use of vaccine. Flare-ups of epidemics in an immune population can be derived to reflect either the occurrence of specific pockets of interacting, nonimmunized individuals or of a general insufficient coverage of vaccination. Again, it should be emphasized that there is no evidence of spread of measles in the form of contageous subclinical infections in an extensively vaccinated population. This is reflected in the empiric observations that measles virus lacks capacity to "hibernate" in populations encompassing less than 300 000 people after a regular measles epidemic has burnt out.

Three principle approaches can be taken to tackle the problem of allowing immunization in the presence of maternal antibodies. Candidate products belong in the categories of replicating attenuated measles virus, replicating vectors containing one or more measles virus genes and nonreplicating immunogens including one or more selected components. Each of these categories will be discussed separately.

3.1 Replicating Attenuated Measles Virus

Measles vaccines containing further attenuated strains have been successfully employed for almost 30 years. Their performance is excellent, not the least with regard to induction of long-term immunity preventing the emergence of overt measles upon exposure of vaccinated individuals to wild virus. Situations of discreditation of this kind of vaccine do not reflect inherent deficiencies of these products. Problems in establishing effective herd immunity generally reflect either ineffective vaccination coverage or unacceptable take rates due, for example, in developing countries to the use of inadvertantly thermally inactivated products or to restriction of vaccine virus replication in young individuals with maternal antibodies.

Amplification of attenuation of the vaccine virus may be achieved in two different ways. One possible future approach is to use site directed mutagenesis, but it still remains necessary to develop tools for application of reverse genetics to morbilliviruses. Even if this would be possible, problems would remain in selecting the appropriate targets for genomic alterations, since sufficient knowledge of the molecular pathogenic properties of measles virus is not available. In fact, such knowledge may be long in waiting considering the complexity of the large measles virus genome and the assumed intricate orchestration of genes with individual qualitative uniqueness. A second way to potentially alter the virus might take into consideration the collective influence of linear and discontinuous immunogenic sites in virus neutralization. This approach can be targeted but it is dependent on the availability of appropriate reagents, with respect to neutralizing monoclonal reagents. A critical conceptual issue is the definition of monotypic properties. Is this property dependent on certain critical sites among a larger family of sites interacting with neutralizing antibodies or is it dependent on the aggregated effect of reactions involving a certain fraction of all sites potentially engaged in neutralization?

Only limited studies of neutralization escape mutants have been performed so far, but they may now be expanded by use of recently introduced technology. In the early 1980s several laboratories generated a number of H-specific murine monoclonal antibodies. By use of competition ELISA a number of overlapping antigenic sites were defined. Consecutive passaging in this laboratory of virus in the presence of monoclonal antibodies representing each the of four different sites identified allowed the selection of a measles virus strain which could not be neutralized by any of the available clonal H-specific antibodies (SHESHBERADARAN and NORRBY 1986). These markedly modified escape mutants still were effectively neutralized by a polyclonal hyperimmune serum or by convalescent human sera. In a later study the linear and discontinuous nature of individual sites and their specific amino acid dependence was defined in these mutants (HU et al. 1993) Interestingly some sites mapped to a small region of the H polypeptide which is expected to be exposed since it harbors the four glycosylation sites that are used.

It is not known at this stage how representative the different antigenic sites, defined by use of fortuitously isolated monoclonal antibodies, are for the normal overall neutralizing immune interactions involving the virus. Potentially this can now be examined. It would seem possible to use the randomized library technique for production of human Fab antibody fragments in E. Coli bacteria to generate a wide array of neutralizing measles antibodies (BARBAS et al. 1991; WILLIAMSON et al. 1993). Hopefully this range of activities would be representative of the complex polyclonal immune response after infection. In the process of selection of these antibodies targeting of H- and F-specific antibodies might be made, since critical immune reactions of postinfection antibodies can be inferred to involve predominantly these two virus components. Once a wider range of neutralizing reagents has been produced it remains to derive virus variants that can no longer be neutralized by a collection of the immune reagents by consecutive passaging in the presence of individual clonal antibodies. Eventually the capacity of these variants to be neutralized by polyclonal sera should be determined. Possibly some of these variants are only partially neutralized by such reagents due either to an overall quantitative reduction of the number of neutralizing sites available or to the absence of certain qualitatively unique sites of this kind. The search for immune selection modified viruses could be restricted by two circumstances. One potential limitation could be if it turns out that nonneutralizable variants cannot be generated when particularly unique and critical sites are involved. Another possibly hampering circumstance would be if effects identified by in vitro neutralization assays would turn out to be only a partial representation of in vivo protective, antibody-dependent effects. Be this as it may, the approach is still worth the effort and it is rational with regard to the specific goal set. Since the intention is to use the modified virus as a live vaccine it is obvious that the selection experiments should depart from a vaccine strain of virus. The aim is not to generate a new type of measles virus by sidestepping the multitargeted immune response elicited by infection or by immunization!

From a more speculative vantage point, still another approach using live virus could be conjectured. Morbilliviruses have a remarkably conserved F component, whereas the attachment protein H is highly divergent, possibly reflecting the use of different cellular receptors for various members of the genus (SHESHBERADARAN et al. 1986). It has been experienced that in some cases a heterologous morbillivirus can be used for immunization, as, for example, the effective use of live measles virus as a vaccine for puppies needing protection against canine distemper virus. Implicitly, the protection provided depends predominantly on F component-specific immunity (APPEL et al. 1984). In the evolutionary tree of morbilliviruses the closest relative of measles virus is bovine rinderpest virus. In fact measles virus evolved from the latter virus after it had split from peste de petit ruminant virus (DIALLO et al. 1994). Measles and bovine rinderpest viruses not only have very similar F proteins, but they also display an H component cross-reactivity (SHESHBERADARAN et al. 1986) One therefore wonders what immunizing capacity live rinderpest virus would have in humans. Could this be an attractive candidate for a virus to replicate in the presence of measles-specific maternal antibodies, with the aim of inducing a state of protection against measles infections in young infants? Obviously, if such a virus was to be used in nature, it needs to be disarmed so that it no longer can cause bovine rinderpest.

3.2 Replicating Vectors Containing One or More Measles Virus Genes

Identification of individual measles virus genes by molecular cloning techniques has provided opportunities for expression of one or more of these genes by use of viral or bacterial vectors. This approach has a number of attractive features. Intracellular processing of antigens would allow for generation of both antibody-dependent and cell-mediated immune reactions. If the vector is chosen not to represent a ubiquitous microbe in the human habitat, efficient initial replication of the vector should be possible. Depending upon the vector chosen it may also be possible to use this more than one time. Finally, replication of the selected vector at mucosal membranes may preferentially convey an immunity that restricts replication of measles virus at its portal of entry. However, in the latter situation it is of particular importance to evaluate the durability of immunity established.

Genes to be included in vectors may be selected for two different purposes. One purpose would be to induce a broad immunity including both humoral and cell-mediated effects by expressing one or both genes for the surface glycoproteins H and F. Alternatively, the approach would be to aim exclusively for cell-mediated immune phenomena and hence preferentially only include genes for internal components, in the first place the N gene. The interference by maternal antibodies would be expected to be minimized in the latter case. Both approaches have been taken as mentioned above.

Different vectors have been tested for expression of measles virus genes. All measles virus genes except the L gene have been expressed alone or in combinations by vaccinia virus recombinants and the protective efficacy of these vectors was evaluated in different experimental rodent models (BRINCKMANN et al, 1991; WILD et al. 1992). Originally, it was shown that immunity induced by the surface glycoproteins H and F in BALB/c mice could protect these animals against a fatal infection after intracerebral injection of neurotropic measles virus (DRILLIEN et al. 1988; WILD et al. 1992). In later studies it was demonstrated that a vaccinia virus recombinant carrying only the gene for one of the internal components of measles virus, the N protein, sufficed to provide protection (BANKAMP et al. 1991). This effect was interpreted to reflect establishment of T cell cytotoxic immune reactions. Thus, it appears that when vaccinia virus recombinants carrying different measles virus genes are used there is variable recruitement of humoral and cell-mediated immune functions depending on the nature of the immunogen (BRINKMANN et al. 1991; BEAUVERGER et al. 1993, personal communications; NIEWIESK et al.1993; WILD et al. 1992). It has also been analyzed to what extent measles virus gene-specific recombinants can provide protection against canine distemper virus in mice (WILD et al. 1993) or in dogs (TAYLOR et al.1991b). Heterologous protection was seen after immunization with the F and the N protein and partial protection was found when a vector carrying the M gene was used. Unexpectedly it was found in one of the studies that also the H protein gave some heterologous protection (TAYLOR et al. 1991b). The basis for this protection needs to be further examined.

Heterologous protection of measles immunization against canine distemper was also explored by use of avipox virus instead of vaccinia virus as a vector (TAYLOR et al. 1992). The unique feature of this poxvirus system is its self-containing nature (Baxby and Paoletti 1992). Replication of avipoxvirus is restricted in human cells and only a single incomplete replication of virus occurs after infection. Thus, the efficacy of immunization would be expected to be dependent on the number of cells initially infected and on the amount of measles antigen produced in individual cells. It was demonstrated that the efficacy of immunization did not differ when avipox virus was used instead of vaccinia virus. This unexpected relative efficacy of the former vector could mean either that it causes production of enhancing immunomodulating factors or that it lacks the capacity to induce production of immunization suppressing factors, potentially associated with vaccinia virus replication. The usefulness of avipox vectors carrying measles virus genes for immunization in humans is currently being evaluated.

Another potentially useful viral vector besides poxviruses is adenoviruses which have the capacity to replicate in the respiratory epithelium. Finally, attempts are also being made to exploit bacterial vectors. In this case bacille Calmette Guerin (BCG) has been proposed as one means to enhance early T cell priming for a selected gene product (STOVER et al. 1991), which in the case of measles has been chosen to be the N protein.

There is still one other technique which may be feasible for intracellular production of selected (viral) immunogens; this is DNA immunization (TANG et al. 1992). Available technology would allow evaluation of the immunizing effect of intracellularly produced, e.g., measles virus H and F proteins by appropriate deposition of cDNA representing these products. Such a deposition may be made in muscle or epithelial tissues, with potentially variable effects on the durability of the immunity produced.

3.3 Nonreplicating Immunogens

Already in the early 1960s attempts were made to develop inactivated measles vaccines. This approach appeared promising since it was known that antibodies provided in the form of pooled immunoglobulin and given parenterally, even in low concentrations, effectively could prevent measles. Thus it was reasoned that induction of stable titers of circulating antibodies by use of inactivated products, which at the time were either treated with formaldehyde or Tween 80 and ether, in an optimized schedule should give efficient protection. These early studies had two problems. One was the difficulty in obtaining stable antibody titers. Aluminum hydroxide seemed to be an inefficient vehicle and multiple repeated injections had to be used. The second problem encountered was more unexpected. As eventually determined in studies performed over many years, it turned out that both the inactivating procedures used had a selective detractive effect on the immunogenicity of the F protein. The protein was not rendered non-immunogenic but the specific antibodies produced did not interfere with the critical fusion activity of the component. Thus from the functional perspective the humoral immune response was unbalanced and preferentially geared towards an anti-H response (NORRBY et al. 1975). Hereby the stage was set for immune pathological reactions. It should be emphasized that the problems encountered related to the particular crude products and inactivating procedures used during the early 1960s. Since then, advanced technology has been developed that allows production of native H and F proteins in large quantities in different systems. There is no reason to anticipate the emergence of any immune pathological complication provided a balanced humoral response towards both surface glycoproteins, mimicking their presence in pooled immunoglobulins, can be generated.

Different approaches can be selected to produce measles virus glycoproteins since the genes directing the synthesis of various products have been isolated and immune reagents specific for each component are available. In essence, the challenge of effectively producing pure measles components is similar to that of preparating other kinds of glycoproteins by modern biotechnological procedures. In its simplest form, highly purified glycoproteins can be isolated from crude tissue culture material by use of immunoaffinity chromatography employing monoclonal antibodies (VARSANYI et al. 1984, 1987; DE VRIES et al. 1988). As an example, the H and F components of the homologous morbillivirus, canine distemper virus, were purified by immunoaffinity chromatography and used to induce component specific immune responses in dogs. Under the conditions of immunization the F component induced a protection that was infection-permissive whereas immunity relying on antibodies to the H-component was complete

(NORRBY et al. 1986). The potential of exploiting primarily the cross-reactive nature of F-specific immunity in using one morbillivirus to protect against the other was already alluded to above.

By use of different vectors, e.g., in the baculovirus system or in the form of the above-mentioned vaccinia virus vectors, a particular enrichment of a selected morbillivirus component in cells can be obtained. Hereby effective procedures for production of large quantities of purified components should be at hand. The measles virus nucleocapsid has been produced in E. Coli and used for mapping of T cell helper epitopes (GIRAUDON et al. 1991). The baculovirus system was used for synthesis of measles virus H and F glycoproteins in SF9 insect cells (VIALARD et al. 1990). The yield was 50-150 mg per 10^8 cells. The glycosylation of the proteins was somewhat impaired and the F0 precursor was not completely cleaved into F1 and F2 proteins. Nonetheless, the viral components showed activity in hemagglutination, hemolysis and fusion assays.

A particular kind of nonreplicating measles vaccine would be a synthetic immunogen including selected peptides representing both B and T cell antigenic sites. Attempts have been made to define critical sites in the envelope glycoproteins (PARTIDOS et al. 1991; PARTIDOS and STEWARD 1992; OBEID et al. 1993; BEAUVERGER et al., personal communication) and in the internal N protein (GIRAUDON et al. 1991; BEAUVERGER et al. 1993). Certain B cell-dependent sites were defined in the F and the H proteins and peptides representing these sites induced antibodies reacting with the peptide and the whole virus protein. However, as yet no antibody activity of biological relevance was demonstrated. Single T cell helper and class I-restricted T cytotoxic sites were also defined in the H protein. It remains to be shown if any combination of peptides representing defined B and T cell sites in measles virus proteins can induce protective immunity in experimental animals.

An as yet unsolved problem is to identify suitable vehicles for presentation of the purified viral antigens. Many different vehicles have been proposed, but it seems that each of them have their shortcomings. Thus there is a need for further work on various delivery systems. Obviously advances in this field will have repercussions not only in the field of measles vaccinology.

If an effective non-replicating measles virus immunogen can be generated, and the problem of delivery can be solved, the appropriate use of such a product in young children remains to be decided. Repeated injections probably have to be given unless some slow release vehicle is employed that allows for generation of both a primary immune response and maturation into a response conventionally obtained by boostering. Conditions of prolonged immunization might be established by use of the abovementioned DNA immunization technique. Multiple contacts with children before the age of 6 months would be needed in an "expanded program of immunization" (EPI), exploiting a new kind of non-replicating measles vaccine that can be given at an early age. Furthermore, an additional immunization most likely needs to be given at the age of 12–18 months, perhaps then preferably with a conventional live measles vaccine. In this case an appropriate balance of levels of immunity produced needs to be adjusted to allow replication of vaccine virus. This may be a difficult task, but the alternative, to base life-long immunity exclusively on the use of nonreplicating immunogens, may also have its shortcomings.

4 Epilogue

The measles vaccines that are in use today were developed during the late 1950s and early 1960s. Effective live vaccines evolved from these endeavours and these vaccines have been successfully adminstered in industrialized and in developing countries. However certain problems in preventing epidemic measles have been encountered. These problems, which mostly are of a practical-logistic nature, need to be solved before global eradication of disease and the ensuing elimination of infection can be achieved. A problem of particular urgency is the yearly death toll due to measles that still occurs in developing countries. The availability of a vaccine that can be used for effective immunization in the presence of moderate amounts of maternal antibodies would markedly facilitate resolution of this problem. Several different kinds of new vaccines may be considered.

As discussed in this chapter, new knowledge about the molecular genetics and immunobiological properties of measles virus and the availability of many new technologies provide excellent opportunities for new advances in the field of measles vaccinology. The selection of preferred approaches need to be anchored in precise definitions of the expected quality of new vaccine product and in classifications of tools available to develop the appropriate product. In the present exploratory phase, both live and inactivated vaccines need to be considered. Among live products the choice will be between further modified whole attenuated virus, recombinant vector constructs or cDNA representing selected genes for intracellular administration. Nonreplicating vaccine products range from whole viral proteins to synthetic immunogens encompassing peptides representing dominant linear B and T cell sites. The use of a nonreplicating immunogen will demand the use of repeated injections. This fact, together with the expected relatively higher cost for a product of this kind, may restrict its application. However, live vaccine products, in particular as represented by whole virus, may have an inherent limitation when given in the presence of maternal antibodies.

Since it is as yet unclear which of the different options for development of new measles vaccines should be given priority, invigorating efforts should be made in furthering the use of already existing live vaccines. It is not impossible that the appropriate use of these products, especially in the form of mass campaigns canvassing large populations, will lead to an elimination of measles prior to the launching of a new kind of measles vaccine.

References

Albrecht P, Hermann K, Burns RG (1981) Role of virus strain in conventional and enhanced plague neutralization test, J Virol Methods 3; 251–260

Appel MJG, Shek WR, Sheshberadaran H, Norrby E (1984). Measles virus and inactivated canine distemper virus induce incomplete immunity to canine distemper. Arch Virol 82: 73–82

Bankamp B, Brinckmann UG, Reich A, Niewiesk S, ter Meulen V, Liebert UG (1991). Measles virus nucleocapsid protein protects rats from encephalitis. J Virol 65: 1965–1700

Barbas III CF, Kang AS,Lerner RA, Benkovic SJ (1991). Assembly of combinatorial antibody libraries on phage surfaces: The gene III site. Proc Natl Acad Sci USA 88; 7978–7982.

Baxby D, Paoletti E (1992) Potential use of non-replicating vectors as recombinant vaccine. Vaccine 10: 8–11

Beauverger P, Buckland R, Wild TF (1993) Measles virus antigens induce both type-specific and canine distemper virus cross-reactive cytotoxic T lymphocytes in mice: localization of a common L^d-restricted nucleoprotein epitope. J Gen Virol 74: 2357–2363

Birrer MJ, Udem S, Nathanson S, Bloom BR (1981) Antigenic variants of measles virus. Nature 293: 67–69

Brinckmann UG, Bankamp B, Reich A, ter Meulen V, Liebert UG (1991) Efficacy of individual measles virus structural proteins in the protection of rats from measles encephalitis. J Gen Virol 72: 2491–2500

Carter MJ, Willcocks MM, Loffler S, ter Meulen V (1982) Relationship between monoclonal antibody-binding sites on the measles hemagglutinin. J Gen Virol 63: 113–120

de Vries P, van Binnendijk RS, van der Marel P, Beuvery EC, van Wezel AL, Voorma HO, UytdeHaag FGCM, Osterhaus ADME (1987) Inhibition of measles virus induced cell-cell fusion with a monoclonal antibody directed against the hemagglutinin. Viral Immunol 1: 25–31

de Vries P, van Binnendijk RS, van der Marel, P, van Wezel AL, Voorma HO, Sundqvist B, UytdeHaag FGCM, Osterhaus ADME (1988) Measles virus fusion protein presented in an immune-stimulating complex(iscom) induces hemolysis-inhibiting and fusion-inhibiting antibodies, virus-specific T cells and protection in mice. J Gen Virol 69: 549–559

Diallo A, Barrett T, Barbron M, Meyer G, Lefevre PC (1994) Cloning of the nucleocapsid protein gene of peste des petits ruminants virus: relationship to other morbilliviruses. J Gen Virol 75: 233–237

Drillien R, Spelmer D, Kirn A, Giraudon P, Buckland R, Wild TF, Lecocq JP (1988). Protection of mice from fatal measles encephalitis by vaccination with vaccinia virus recombinants encoding either the hemagglutinin or the fusion protein. Proc Natl Acad Sci USA 85: 1252–1256

Edmondson M, Addiss D, McPherson J, Berg J, Circo S, Davis J (1990) Mild measles and secondary vaccine failure during a sustained outbreak in a highly vaccinated population. J Am Med Assoc 263; 2467–2471

Giraudon P, Wild TF (1985) Correlation between epitopes on hemagglutinin of measles virus and biological nativities. Passive protection by monoclonal antibodies is related to their hemagglutination inhibiting activity. Virology 144: 46–58

Giraudon P, Jacquier MF, Wild TF (1988) Antigenic analysis of African measles virus field isolates: identification and localization of one conserved and two variable epitope sites on the NP protein. Virus Res 18: 137–152

Giraudon P, Buckland R, Wild TF (1991) The immune response to measles virus in mice. T-helper response to the nucleoprotein and mapping of T-helper epitopes. Virus Res 22: 41–54

Hu A, Sheshberadaran H, Norrby E, Kovamees J (1993) Molecular characterization of epitopes on the measles virus hemagglutinin protein. Virology 192: 351–354

Niewiesk S, Brinckmann U, Bankamp B, Sirak S, Liebert UG, ter Meulen V (1993). Susceptibility to measles virus-induced encephalitis in mice correlates with impaired antigen presentation to cytotoxic T lymphocytes. J Virol 67: 75–81

Norrby E, Gollmar Y, (1975) Identification of measles virus-specific hemolysis-inhibiting antibodies seperate from hemagglutination-inhibiting antibodies. Infect Immun 11: 231–239

Norrby E, Enders-Ruckle G, ter Meulen V (1975) Differences in the appearance of antibodies to structural components of measles virus after immunization with inactivated and live virus. J Infect Dis 132: 262–269

Norrby E, Utter G, Orvell C, Appel M (1986) Protection against canine distemper virus in dogs after immunization with isolated fusion protein. J Virol 58: 536–541

Obeid OE, Partidos CD, Steward MW (1993) Identification of helper T-cell antigenic sites in mice from the haemagglutinin glycoprotein of measles virus. J Gen Virol 74: 2549–2557

Partidos CD, Stanley CM, Steward MW (1991) Immune responses in mice following immunization with

chimeric synthetic peptides representing B and T cell epitopes of measles virus proteins. J Gen Virol 72: 1293–1299.
Partidos CD, Steward MW (1992) The effects of a flanking sequence on the immune response to a B and a T cell epitope from the fusion protein of measles virus. J Gen Virol 73: 1987–1994
Rota JS, Hummel KB, Rota PA, Bellini WJ (1992) Genetic variability of the glycoprotein genes of current wild-type measles isolates, Virology 188: 135–142
Sato TA, Fukuda A, Sugiura A (1985) Characterization of major structural proteins of measles virus with monoclonal antibodies. J Gen Virol 66: 1397–1409
Sheshberadaran H, Chen S-H, Norrby E (1983) Monoclonal antibodies against five structural components of measles virus. 1. Characterization of antigenic determinants on nine strains of measles virus. Virology 128: 341–353
Sheshberadaran H, Norrby E, McCullough KC, Carpenter WC, Orvell C (1986) The antigenic relationship between measles, canine distemper and rinderpest viruses studied with monoclonal antibodies. J Gen Virol 67: 1381–1392
Sheshberadaran H, Norrby E (1986) Characterization of epitopes on the measles virus hemagglutinin. Virology 152: 58–65
Stover CK, de la Cruz VF, Fuerst TR, Burlein JE, Benson LA, Bennett LT, Banal GP, Young JF, Lee MH, Hatfull G-F, Snapper S-B, Barletta R, Jacobs WR, Bloom BR (1991) New use of BCG for recombinant vaccines. Nature 351: 456–460
Tang DC, DeVit M, Johnson SA (1992) Genetic immunization is a single method for eliciting an immune response. Nature 356: 152–154
Taylor MJ, Godfrey E, Baczko K, ter Meulen V, Wild TF, Rima B (1991a) Identification of several different lineages of measles virus. J Gen Virol 72: 83–88
Taylor J, Pincus S, Tartaglia J, Richardson C, Alkhatib G, Briedis D, Appel M, Norton E, Paoletti E (1991b) Vaccinia virus recombinants expressing either the measles virus fusion or hemagglutinin glycoprotein protect dogs against canine distemper virus challenge. J Virol 65: 4263–4274
Taylor J, Weinberg R, Tartaglia J, Richardson C, Alkhatib G, Briedis D, Appel M, Norton E, Paoletti E (1992) Nonreplicating viral vectors as potential vaccines: Recombinant canarypox virus expressing measles virus fusion (F) and hemagglutinin(HA) glycoproteins. Virology 187: 321–328
van Binnendijk RS, van Baaklen CA, Poelen MCM, de Vries P, Boes J, Cerundolo V, Osterhaus ADME, UytdeHaag FGCM (1992) Measles virus transmembrane fusion protein synthesized de novo or presented in iscom is endogenously processed for HLA class I and II restricted cytotoxic T cell recognition. J Exp Med 176: 119–129
Varsanyi TM, Utter G, Norrby E (1984) Purification, morphology and antigenic characterization of measles virus envelope components. J Gen Virol 65: 355–366
Varsanyi TM, Morein B, Love A, Norrby E (1987) Protection against lethal measles virus infection in mice by immune stimulating complexes containing the hemagglutinin or fusion protein. J Virol 61: 3896–3901
Vialard J, Lalumiere M, Vernet T, Briedis D, Alkhatib G, Hemming D, Levin D, Richardson C (1990) Synthesis of the membrane fusion and hemagglutinin proteins of measles virus using a novel baculovirus vector containing the B-galactosidase gene. J Virol 64: 37–50
Wild TF, Malvoisin E, Buckland R (1991) Measles virus: both the hemagglutinin och fusion glycoproteins are required for fusion. J Gen Virol 72: 439–442
Wild TF, Bernard A, Spelmer D, Drillien R (1992) Construction of vaccinia virus recombinants expressing several measles virus proteins and analysis of their efficacy in vaccination of mice. J Gen Virol 73: 359–367
Wild TF, Bernard A, Spelmer D, Villeval D, Drillien R (1993) Vaccination of mice against canine distemper virus-induced encephalitis with vaccinia virus recombinants encoding measles or canine distemper virus antigens. Vaccine 11: 438–444
Williamson RA, Burioni R, Sanna PP, Partridge LJ, Barbas CF III, Burton DR (1993) Human monoclonal antibodies against a plethora of viral pathogens from single combinatorial libraries. Proc Natl Acad Sci USA 90: 4141–4145

Appendix: Measles Virus Antigenome and Protein Consensus Sequences

F. RADECKE and M.A. BILLETER

During the last few years, many new sequence analyses have been reported, primarily for circulating measles virus (MV) strains. Majority consensus sequences of the antigenome and of the proteins were established using all wild-type and vaccine strain sequences available from GenEMBL on June 1, 1994, with additional wild-type sequences for the H gene and the L gene (kindly provided by K. Baczko) and wild-type and vaccine strain sequences of the 5'- and 3'-terminal 100 nucleotides (kindly provided by S. Udem). Note that no sequences derived from subacute sclerosing panencephalitis (SSPE) or measles inclusion body encephalitis (MIBE) have been considered.

The alignments were made in the following groupings using the DNASTAR MegAlign-program: leader and N 5'UTR (nt 1-107; 15 sequences), N ORF (nt 108-1685; 39 sequences of which 11 are carboxyl-terminal, starting at nt 1230), N 3'UTR (nt 1686-1744; 6 sequences), P gene (nt 1748-3402; 7 sequences), M 5'UTR (nt 3406-3437; 7 sequences), M ORF (nt 3438-4445; 18 sequences), M 3'UTR (nt 4446-4871; 7 sequences), F 5'UTR (nt 4875-5448; 5 sequences), F ORF (nt 5449-7110; 16 sequences), F 3'UTR (nt 7111-7247; 7 sequences), H gene (nt 7251-9208; 33 sequences), L 5'UTR and ORF (nt 9212-15785; 7 sequences), and L 3'UTR and trailer (nt 15786-15894; 18 sequences). These majority consensus sequences replace the appropriate regions in the sequence of the MV strain Edmonston (GenEMBL accession number K01711). Figure A1 shows a schematic representation of the MV antigenome and its ORFs. The proposed total length is 15894 nt. It might be relevant that this number conforms to the "Rule of Six" (CALAIN and ROUX 1993). Figure A2 gives the majority consensus sequences for the antigenome (as cDNA) and for the proteins.

Upper case letters represent residues which are 100% conserved in all reported sequences. Lower case letters indicate positions where deviations have been found (percentage of conservation ranging from 50% to < 100%). Note that some of these deviations possibly constitute sequencing errors.

Nucleotides underlined in boxes A and B in the 5'-terminal region of the antigenome are complementary to those in boxes A' and B' near the 3' terminus.

Δ below a small letter marks residues which are deleted in at least one reported sequence. (Δ) denotes residues found to be either deleted or replaced. Arrow heads indicate the positions where given nucleotides were reported to be inserted.

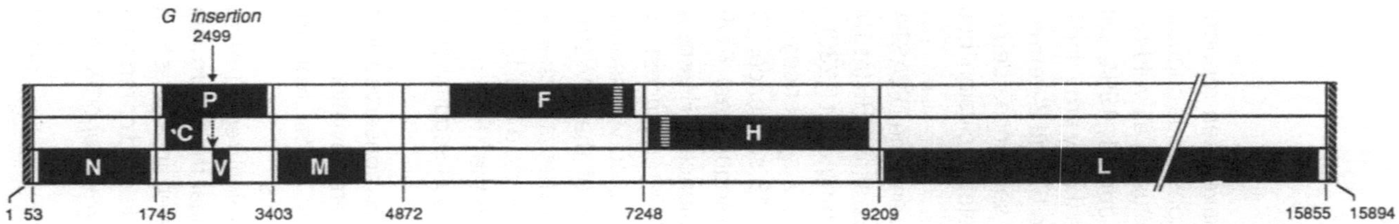

Symbols

- = leader
- = untranslated region (UTR)
- = gene boundary trinucleotide
- = open reading frame (ORF)
- = change of ORF
- = transmembrane coding region
- = trailer
- 53 = position of the first residues of the gene boundary trinucleotides

Gene numbering (nt)
(numbers in brackets: counted from transcription start)

	Transcription		Translation	
	Initiation	Termination	Start (first nt of start codon)	Stop (last nt of stop codon)
N gene	56 (1)	1744 (1689)	108 (53)	1685 (1630)
P gene	1748 (1)	3402 (1655)	1807 (60)	3330 (1583)
C			1829 (82)	2389 (642)
V			1807 (60)	2705 (958)
M gene	3406 (1)	4871 (1466)	3438 (33)	4445 (1040)
F gene	4875 (1)	7247 (2373)	5449 (575)	7110 (2236)
H gene	7251 (1)	9208 (1958)	7271 (21)	9124 (1874)
L gene	9212 (1)	15854 (6643)	9234 (23)	15785 (6574)

Fig. A1. Schematic representation of the measles virus antigenome and its open reading frames

```
leader
ACCAAACAAAGTTGGGTAAGGATAGtTCAATCAATGATCATcTTCTAGTgCACTTAGGATTCAAGATCCTATTaTCAGGGACAAGAgCAGGAtTAgGGAT   100
   A box                                                                                  B box

      N protein
ATCCGAGATGGCCACACTttTaAGGAGCTTAGCAtTGTTCAAAAGaAaCAAGGACAAACCACCCATtACATCAGGATCCGGTGGAGCCATCAGAGGAATC   200
       M  A  T  L  L  R  S  L  A  L  F  K  R  n  K  D  K  P  P  I  T  S  G  S  G  G  A  I  R  G  I

AAACACATTATTATAGTACCAATcCCtGGAGATTCCTCAATTACCACTCGATCcAGACTtCTGGACcGGTTGGTcAGgTtAATTGGAaACCCGGATgTgA   300
 K  H  I  I  I  V  P  I  P  G  D  S  S  I  T  T  R  S  R  L  L  D  R  L  V  R  I  I  G  n  P  D  v

GCGGGCCCAAAcTAACAGGgGCACTAATAGGTATAtTaTCCtTATTTGTGGAGTCTCCAGGTCAATTGATTCAgAGGATCACcGATGACCCTGACGTTAG   400
S  G  P  K  L  T  G  A  L  I  G  I  L  S  L  F  V  E  S  P  G  Q  L  I  Q  R  I  T  D  D  P  D  V  S

CATaAGGCTGTTaGAGGTTGTcCAGAGtGACCAGTCACAATCTGGCCTtACCTTcGCATCAAGAGGTACcAAcATGGAGGATGAGGCGGAccAaTAcTTT   500
  I  R  L  L  E  V  V  Q  S  D  Q  S  Q  S  G  L  T  F  A  S  R  G  T  N  M  E  D  E  A  D  q  Y  F

TcaCATGATGATcCAAttAGTaGTGATCAATCCAGGTtcGGaTGGTTcGaGAaCAAGGAAATCTCAGATATTGAAGTGCAAGAcCCTGAGGGATTCAaCA   600
 s  H  D  D  p  I  S  s  D  Q  S  R  f  G  W  F  e  n  K  E  I  S  D  I  E  V  Q  D  P  E  G  F  n

TGATTCTGGGtACCATcCTaGCcCAaATtTGgGTCTTGcTCGCAAAGGCgGTtACGGCcCCAGACACgGCAGCtGAtTCGGAGCTAAGAAGGTGGATAAA   700
M  I  L  G  T  I  L  A  Q  I  w  V  L  I  A  K  A  V  T  A  P  D  T  A  A  D  S  E  L  R  R  W  I  K

gTACACCCAACAAAGAAGgGTAGTTGGTGAATTTAGaTTGGAGAGAAaATGGTTGGATGTGGTGAGgAAcAggATTGCCGAgGACCTCTCcTTACGCCGa   800
  Y  T  Q  Q  R  R  V  V  G  E  F  R  L  E  R  k  W  L  D  V  V  R  N  r  I  A  E  D  L  S  L  R  R

TTCATGGTcGCtCTAATCCTGGATATCAAGAGaACACCCGGaAAcAAACCcAGGATTGCTGAAATGATATGTgACATTGATACATATATCGTAGAGGCAG   900
 F  M  V  A  L  I  L  D  I  K  R  T  P  G  N  K  P  R  I  A  E  M  I  C  d  I  D  T  Y  I  V  E  A

GAtTAGCCAGTTTTATCCTgACTATTAAGTTtGGgATAGAAACtATGTAtCCTGCTCTTGGAcTGCATGAATTTGctGGTGAGTTATCCACACTTGAgTC  1000
G  L  A  S  F  I  L  T  I  K  F  G  I  E  T  M  Y  P  A  L  G  L  H  E  F  a  G  E  L  S  T  L  E  S

CTTGATGAAcCTTTACCAgCAAATGGGggAAaCTGCACCcTACATGGTAAtcCTgGAGAACTCAATTCAGAACAAgTTCAGtGCAGGAtCATACCCTCTG  1100
  L  M  N  L  Y  Q  Q  M  G  e  t  A  P  Y  M  V  I  L  E  N  S  I  Q  N  K  F  S  A  G  s  Y  P  L

CTCTGGAGCTATGCCATGGGAGTAGGAGTGGAACTTGAAAACTCCATGGGAGGTTTGAACTTtGGcCGaTCTTACTTtGATCCAGCATAtTTtAgAtTAG  1200
 L  W  S  Y  A  M  G  V  G  V  E  L  E  N  S  M  G  G  L  N  F  G  R  S  Y  F  D  P  A  Y  F  r  L

GGCAAGAGATGGTaAGGAGgTCAGCTGGAAAGGTCAGtTCCACATTgGCATCtGaACTcGGtATCACtGCCGAgGAtGCaAGGCTTGTtTCAGAGAttGC  1300
G  Q  E  M  V  R  R  S  A  G  K  V  S  S  T  L  A  S  e  L  G  I  T  A  E  D  A  R  L  V  S  E  I  A

AATGCAtACtaCtGAGGACAgGATCAGtAGAGCgGTtGGaCCcAGaCAagCCCAAGTaTCATTtcTACAcGGTGATCAAAgTGAgAAtGAGCTaCCgaGA  1400
  M  H  T  t  E  D  r  I  S  R  A  V  G  P  R  Q  a  Q  V  S  F  I  h  G  D  Q  s  E  N  E  L  P  r

ttGGGGGgcAAGGAaGAtAggAGgGtCAAACAgAGtCGaGGAGAAgCcaGgGAGAgCTACAGAGAaaCcgGgcCCAGcAGAgCAaGTGAtGCgAGAgctg  1500
 I  G  g  K  E  D  r  R  v  K  Q  S  R  G  E  a  r  E  s  Y  R  e  t  g  p  S  R  a  s  d  A  R  a

CCCAtCtTCCaACCggCAcaCcCcTAGaCATTGACACTGcAtcgGAGTccaGCCaAGAtCcgCAGGAcAGTCgAAGGTCaGCTGAcgCccTgCTtAGGcT  1600
a  H  I  P  T  g  t  p  L  d  I  D  T  a  s  E  s  s  q  D  p  Q  D  S  r  R  S  A  d  a  L  L  R  L

GCAaGCCATGGCAGGAATCTcGGAAGAACAAGGcTCAGAcAcGGACAcCCcTagaGTgTACAATGACAgAgATCTtCTAGACTAgGTGCgAGAGGCCGAG  1700
  Q  A  M  A  G  I  s  E  E  Q  G  S  D  t  D  t  p  r  V  Y  N  D  r  d  L  L  D  *

GaCCAGAACAACATCCGCCTACCCTCCATCATTGTTATAAAAAACTTAGGAACCAGGTCCACACAGCCGCCAGCCCATCAACCATCCACTCCCACGATTG  1800
```

Fig. A2. The majority consensus sequences for the measles virus antigenome (as cDNA) and for the proteins (Continued on pp 184–191)

```
     P/V protein          C protein
GaGCCgATGGCAGAAGAGCAGGCACGCCATGTCAAAAACGGACTGGAATGCATCCGGGCTCTCAAGGCCGAGCCCATCGGCTCACtGGCCATCGAGGAAG  1900
      M  A  E  E  Q  A  R  H  V  K  N  G  L  E  C  I  R  A  L  K  A  E  P  I  G  S  I  A  I  E  E
                             M  S  K  T  D  W  N  A  S  G  L  S  R  P  S  P  S  A  H  w  P  S  R  K

CTATGGCAGCATGGTCAGAAATATCAGACAACCcAGGACAGGAGCGAGCCaCCTGCAgGGAAGAGAAGGCAgGCAGTtCGGGTCTCAGCAAACCATGCCT  2000
A  M  A  A  W  S  E  I  S  D  N  p  G  Q  E  R  A  t  C  r  E  E  K  A  g  S  s  G  L  S  K  P  C  L
 L  W  Q  H  G  Q  K  Y  Q  T  T  Q  D  R  S  E  P  P  A  g  K  R  R  Q  A  V  R  V  S  A  N  H  A

CTCAGCAATTGGATCAACTGAAGGCGGTGCACCTCGCATCCGCGGTCAGGGAcCTGGAGAGAGCGATGACGACGCTGAAACtTTGGGAATCCCCcCAaGA  2100
  S  A  I  G  S  T  E  G  G  A  P  R  I  R  G  Q  G  p  G  E  S  D  D  D  A  E  T  L  G  I  P  p  r
S  Q  Q  L  D  Q  L  K  A  V  H  L  A  S  A  V  R  D  L  E  R  A  M  T  T  L  K  I  W  E  S  P  Q  E

aATCTCCAGGCATCAAGCACTGGGtTACAGTGTtATTATGTTTATGATCACAGCGGTGAAGCGGTTAAGGGAaTCCAAGATGCTGACTCTATCATGGTTC  2200
 n  L  Q  A  S  S  T  G  L  Q  C  y  Y  V  Y  D  H  S  G  E  A  V  K  G  I  Q  D  A  D  S  I  M  V
  I  S  R  H  Q  A  L  G  Y  S  V  I  M  F  M  I  T  A  V  K  R  L  R  E  S  K  M  L  T  L  S  W  F

AATCAGGCCTTGATGGTGATAGCACCCTcTCAGGAGGAGACAATGAATCTGAAAACAGCGATGTGGATATTGGCGAACCTGATACCGAGGGATATGCTAT  2300
Q  S  G  L  D  G  D  S  T  L  S  G  G  D  N  E  S  E  N  S  D  V  D  I  G  E  P  D  T  E  G  Y  A  I
 N  Q  A  L  M  V  I  A  P  s  Q  E  E  T  M  N  L  K  T  A  M  W  I  L  A  N  L  I  P  R  D  M  L

CACTGACCGGGGATCTGCTCCCATCTCTATGGGGTTCAGGGCTTCTGATGTTGAAACTGCAGAAGGAGGGGAGATCCACGAGCTCCTGAGACTCCAATCC  2400
  T  D  R  G  S  A  P  I  S  M  G  F  R  A  S  D  V  E  T  A  E  G  G  E  I  H  E  L  L  R  L  Q  S
S  L  T  G  D  L  L  P  S  L  W  G  S  G  L  L  M  L  K  L  Q  K  E  G  R  S  T  S  S  *
                                                                                     G insertion
AGAGGCAACAACTTTCCgAAGCTTgGGAAAACTCTCAATGTTCCtCCGCCcCCGGACCCcGGTAGGGCCAGCACTTCCGgGACACCCATTAAAAAGGGCA  2500
 R  G  N  N  F  P  K  L  g  K  T  L  N  V  P  P  P  P  D  P  G  R  A  S  T  S  g  T  P  I  K  K  G
                                                                     V protein C-terminal region  └→H
CAGAcgcgAGATTAGCCTCATTTGGAACGGaGATCGCGTCTTtATTGACAGGTGGTGCAACCCAATGTGCTCGAAAGTCACCCTCGGAACCATCAGGGCC  2600
T  d  a  R  L  A  S  F  G  T  e  I  A  S  I  L  T  G  G  A  T  Q  C  A  R  K  S  P  S  E  P  S  G  P
 R  r  e  I  S  L  I  W  N  G  D  R  V  F  I  D  R  W (C) N  P  M (C) S  K  V  T  L  G  T  I  R  A

AGGTGCACCTGCGGGGAAtGTCCCCGAGTgTGtGAGCAATGCCGcACTGATACAGGAGTGGACACCCGAATCTGGtACCACAATCTCCCCGAGATCCCAG  2700
  G  A  P  A  G  N  V  P  E  c  v  S  N  A  a  L  I  Q  E  W  T  P  E  S  G  T  T  I  S  P  R  S. Q
 R (C) T (C) G  E (c) P  R  V (C) E  Q (C) R  T  D  T  G  V  D  T  R  I  W  y  H  N  L  P  E  I  P

AATAATGAAGAAGGGGGAGActATTATGATGATGAGCTGTTCTCTGATGTCCAAGATATtAAAACAGCCTTGGCCAAAATACACGAGGATAATCAGAAGA  2800
 N  N  E  E  G  G  D  y  Y  D  D  E  L  F  S  D  V  Q  D  I  K  T  A  L  A  K  I  H  E  D  N  Q  K
E  *

TAATCtCCAAGCTAGAaTCACTGCTGTTATTGAAGGGaGAAGTTGAGTCAATtAAGAAGCAGATCAACAGGCAAAATATCAGCATATCCACCcTGGAAGG  2900
I  I  s  K  L  E  S  L  L  L  L  K  G  E  V  E  S  I  K  K  Q  I  N  R  Q  N  I  S  I  S  T  L  E  G

ACACCTcTCAAGCATCATGATCGCCATTCCTGGACTTGGGAAGGATCCCAACGACCCCACTGCAGATGTCGAAATCAATCCCGACTTGAAACCCATCATA  3000
  H  L  S  S  I  M  I  A  I  P  G  L  G  K  D  P  N  D  P  T  A  D  V  E  I  N  P  D  L  K  P  I  I

GGCAGaGATTCAGGcCGAGCACTGGCcGAAGTTCTCAAGAAACCCGtTGCCAGCCGACAAcTCCAAGGAATGACAAATGGACGgACCAGTTCCAGAGGAC  3100
 G  R  D  S  G  R  A  L  A  E  V  L  K  K  P  v  A  S  R  Q  I  Q  G  M  T  N  G  R  T  S  S  R  G

AGCTGCTGAaGgAATTTCAGCtAAAGCCgATCGGGAAAAAGATGAGCTCAGCCGTCGGGTTTGTTCCtGACACCGGCCCTGCATCACGCAGTGTAAtcCG  3200
Q  L  L  k  e  F  Q  I  K  P  I  G  K  K  M  S  S  A  V  G  F  V  P  D  T  G  P  A  S  R  S  V  I  R

CTCCATtATAAAATCcAGCCGGCTAGAGGAGGATCGGAAgCGTTACCTGATGACTCTCCTTGATGAtATCAAAGGAGCCAAtGATCTTGCCAAGTTCCAC  3300
  S  I  I  K  S  S  R  L  E  E  D  R  K  R  Y  L  M  T  L  L  D  D  I  K  G  A  N  D  L  A  K  F  H

CAGATGCTGATGAAGATAATAATGAAGTAGCTACAGCTCAACTtACCTGCCAACCCCATGCCAGTCGACccAaCTAgTACAACCTAAATCCATTATAAAA  3400
 Q  M  L  M  K  I  I  M  K  *                                         (Δ)
                                   M protein
AACTTAGGAGCAAAGTGATTGCCTCCCAAGTTCCACAATGACAGAGATCTACGACTTCGACAAGTCgGCATGGGACATCAAaGGGtCGATCGCTCCGATA  3500
                                     M  T  E  I  Y  D  F  D  K  S  A  W  D  I  K  G  s  I  A  P  I

CAACCcACCACCTACAGTGATGGCAGGCTGGTGCCCCAGGTcAGAGTCATAGATCCTGGTCTAGGcGAcAGgAAgGATGAATGcTttAtGTACATGTtTC  3600
 Q  P  T  T  Y  S  D  G  R  L  V  P  Q  V  R  V  I  D  P  G  L  G  D  R  K  D  E  C  f  m  Y  M  f
```

Fig. A2. (Contd.)

```
TGcTGGGGGTTGTtGaGGACAGcGATccCCTAGGGCCTCCAATCGGGCGaGCATTtGGGTccCTgCcCtTAGGTGTTGGcAGaTCCACAGCAAAgCCCGA  3700
L  L  G  V  V  e  D  s  D  p  L  G  P  P  I  G  R  A  F  G  s  L  p  L  G  V  G  R  S  T  A  k  P  E

agAACTCCTCAAAGAGGCCACTGAGCTtGACAtAGTTGTtAGACGTACAGCAGGGCTCAATGAAAAACTGGTGTTCTACAACAACACCCCACTAACtCTc  3800
 e  L  L  K  E  A  T  E  L  D  I  V  V  R  R  T  A  G  L  N  E  K  L  V  F  Y  N  N  T  P  L  T  L

CTCAcACCTTGGAGAAAGGTCCTAACAACAGGGAGTGTCTTCAAcGCAAACCAaGTGTGCAaTGCGGTTAATCTgATACCGCTcGAtACCcCGCAGAGGT  3900
 L  t  P  W  R  K  V  L  T  T  G  S  V  F  N  A  N  Q  V  C  n  A  V  N  L  I  P  L  D  T  p  Q  R

TCCGTGTTGTtTATATGAGCATCACCCGtCTTTCgGATAACGGGTATTACACCGTTCCtaGAAGAATGCTGGAATTCAGATCGGTCAATGCAGTGGCCTT  4000
F  R  V  V  Y  M  S  I  T  R  L  S  D  N  G  Y  Y  T  V  P  r  R  M  L  E  F  R  S  V  N  A  V  A  F

CAACcTGCTgGTGaCcCTtagGATTGACAAGGCGATagGCCcTGGGAAGATCATCGACAATaCAGAGCAACTTCCTGAgGCAACATTTATGGTCCACATC  4100
 N  L  L  V  t  L  r  I  D  K  A  I  g  p  G  K  I  I  D  N  t  E  Q  L  P  e  A  T  F  M  V  H  I

GGGAACTTCAGGAGAAAGAAGAGTGAAGTCTACTCTGCcGATTATTGCAAAATGAAAATCGAAAAGATGGGCCTGGTTtTTgCACTTGGTGGGATAGGGG  4200
 G  N  F  R  R  K  K  S  E  V  Y  S  A  D  Y  C  K  M  K  I  E  K  M  G  L  V  f  a  L  G  G  I  G

GcACCAGTCTtCACATTAGAAGcACAGGCAAaATGAGCAAGACTCTCCATGCACAACTCGGGTTCAAGAAGACCTTATGTTACCCgCTGATggatATCAA  4300
G  T  S  L  H  I  R  S  T  G  K  M  S  K  T  L  H  A  Q  L  G  F  K  K  T  L  C  Y  P  L  m  d  I  N

TGAaGACCTTAATCGaTTACTCTGGAGGAGCAGATGCAAGATAGTAAGAATCCAGGCAGTTTTGCAGCCATCAGTTCCTCAAGAATTCCGCATTTACGAC  4400
 E  D  L  N  R  L  L  W  R  S  R  C  K  I  V  R  I  Q  A  V  L  Q  P  S  V  P  Q  E  F  R  I  Y  D

GACGTgATCATAAATGATGAcCAAGGACcTATTCAAaGTTCTG[TAG]ACCGTaGTGCCCAGCAATGCCcGAAAACGAcccccctcACAATGACAGCCAGAAG  4500
 D  V  I  I  N  D  D  Q  G  L  F  K  V  L  *                                      Δ

GCCCGGACaAAAAAGCCCCCtCCGAAAgACtCCACgGACCAaGCGaGAGGCCAGCCAGCAGCCGACGGCAAGCgCGAACACCAGGCGGCCCCAGCACAGA  4600

ACAGCCCtGAcACAAGGCCACCACCAGCCAcCCCAATCTGCATCCTCCTCGTGGGACCCCcGAGGACCAACCCCCAAGGcTGCCCCCGAtCCAAaCCACC  4700

AACCGCAtCCCCACCACCCCCGGGAAaGAAaCCCCAGCAAtTGGAAGGCCCCTccCCCTcTtCCTCAACACAAGAACTCCACAACCGAACCGcACAAGC  4800

                  c
GAcCGAGGTGACCCAACCGCagGCAtCCgACTcCCTAGACAGATCCtCTctCCCCGGcAAACT[AAACAAAA][CTT][AGGGCCAAGG]AACATACACACCCAAC  4900
                    Δ

AGAACCCAGACCCCGGCCCACGGCGCCGCGCCCCCAACCCCCGACAACCAGAGGGAGCCCCCAACCAATCCcGCCGGCTCCCCCGGTGCCCACAGGCAGG  5000
                                                                      Δ

                                                                         a
GACACCAACCCCCGAACAGACCCAGCACCCAACCATCGACAATCCAAGACGGGGGGGCCCCCCCAAAAAAAGGCCCCCAGGGGCCGACAGCCAGCACCGC  5100

GAGGAAGCCCACCCACCCCACACACGACCACGGCAACCAAACCAGAACCCAGACCACCCTGGgcCACCAGCTCCcagaCTCGGcCATCACCCCGCAGAAA  5200
                                                              Δ

GGAAAGGCcACAACCCGCGcACCCCAGCCCCGATCCGGCGGGGAGCCACCCAACCCGAACCAGCACCCAAGAGCGATCCCCGAAGGACCCCCGAACCGCA  5300
        Δ          Δ

                                                                                  cacac
AAGGACATCAGTATCCCACAGCCTCTCCAAGTCCCCCGGTCTCCTCCTCTTCTCGAAGGGACCAAAAGATCAATCCACCACACCCGACGACACTCAACTC  5400
```

Fig. A2. (Contd.)

F protein

```
CCCACCCCTAAAGGAGACACCGGGAATCCCAGAATCAAGACTCATCCAaTGTCCATCATGGGTCTCAAGGTGAACGTCTcTGCCATATTCATGGCAGTAC  5500
                                                 m  s  I  M  G  L  K  V  N  V  s  A  I  F  M  A  V

TGTTAACTCTCCAaACACCCACCGGTCAAATCCATTGGGGCAATCTCTCTAAGATAGGGGTGGTAGGaATAGGAAGTGCAAGCTACAAAGTTATGACTCG  5600
 L  L  T  L  Q  T  P  T  G  Q  I  H  W  G  N  L  S  K  I  G  V  V  G  I  G  S  A  S  Y  K  V  M  T  R
                                  ↑ F0

TTCCAGCCATCAaTCATTaGTCATAAAATTAATGCCCAATATAACTCTCCTCAATAACTGCACGAGGGTAGAGATTGCAGAATACAGGAGACTACTGAGA  5700
 S  S  H  q  S  L  V  I  K  L  M  P  N  I  T  L  L  N  N  C  T  R  V  E  I  A  E  Y  R  R  L  L  R

ACaGTTTTGGAACCAATTAGAGATGCACTTAATGCAATGACCCAGAATATAAGACCGGTTCAGAGTGTAGCTTCAaGTAGGAGACACAAGAGATTTGCGG  5800
 T  V  L  E  P  I  R  D  A  L  N  A  M  T  Q  N  I  R  P  V  Q  S  V  A  S  s  R  R  H  K  R  F  A
                                                                                       ···F2 ↑ F1···

GAGTAGTCCTGGCAGGTGCGGCCCTAGGCGTTGCCACAGCTGCTCAGATAACAGCCGGCATTGCACTTCACCAGTCCATGCTGAACTCTCAAGCCATCGA  5900
 G  V  V  L  A  G  A  A  L  G  V  A  T  A  A  Q  I  T  A  G  I  A  L  H  Q  S  M  L  N  S  Q  A  I  D

CAATCTGAGAGCgAGCCTGGAAACTACTAATCAGGCAATTGAGgCAATCaGACAAGCAGGGCAGGAGATGATATTGGCTGTTCAGGGTGTCCAAGACTAC  6000
 N  L  R  A  S  L  E  T  T  N  Q  A  I  E  a  I  r  Q  A  G  Q  E  M  I  L  A  V  Q  G  V  Q  D  Y

ATCAATAATGAGCTGATACCGTCTATGAACCAACTATCTTGTGATTTAATCGGCCAGAAGCTcGGGCTCAAATTGCTCAGATACTATACAGAAATCCTGT  6100
 I  N  N  E  L  I  P  S  M  N  Q  L  S  C  D  L  I  G  Q  K  L  G  L  K  L  L  R  Y  Y  T  E  I  L

CaTTATTTGGCCCCAGcTTACGGGACCCCATATCTGCGGAGATATCtATCCAGGCTTTGAGcTATGCGCTTGGaGGAGAcATCAATAAGGTgTTAGAAAA  6200
 S  L  F  G  P  S  L  R  D  P  I  S  A  E  I  S  I  Q  A  L  s  Y  A  L  G  G  D  I  N  K  V  L  E  K

GCTCGGATACAGTGGAGGTGATTTACTGGGCATCTTAGAGAGCaGAGGAATAAAGGCCCGGATAACTCACGTCGACACAGAGTCCTACTTcATTGTCCTC  6300
 L  G  Y  S  G  G  D  L  L  G  I  L  E  S  r  G  I  K  A  R  I  T  H  V  D  T  E  S  Y  f  I  V  L

AGTATAGcCTATCCGACGCTGTCCGAGATTAAGGGGGTGATTGTCCACCGGCTAGAGGGGGTCTCGTACAAcATAGGCTCTCAAGAGTGGTATACCACTG  6400
 S  I  a  Y  P  T  L  S  E  I  K  G  V  I  V  H  R  L  E  G  V  S  Y  N  I  G  S  Q  E  W  Y  T  T

TGCCCAAGTATGTtGCAACCCAAGGGTACCTTATCTCGAATTTTGATGAGTCATCGTGTACTTTCATGCCAGAGgGgACTGTGTGCAGCCAAAATGCCTT  6500
 V  P  K  Y  V  A  T  Q  G  Y  L  I  S  N  F  D  E  S  S  C  T  F  M  P  E  g  T  V  C  S  Q  N  A  L

GTACCCGATGAGTCCTCTGCTCCAAGAATGCCTCCGGGGGTcCACcAAGTCCTGTGCTCGTACACTCGTATCCGGGTCTTTTGGGAACCGGTTCATTTTA  6600
 Y  P  M  S  P  L  L  Q  E  C  L  R  G  s  T  K  S  C  A  R  T  L  V  S  G  S  F  G  N  R  F  I  L

TCACAAGGGAACCTAATAGCCAATTGTGCATCAATCCTtTGCAAGTGTtACACAACAGGAACGATCATTAATCAAGACCCTGACAAGATCCTAACATACA  6700
 S  Q  G  N  L  I  A  N  C  A  S  I  L  C  K  C  y  T  T  G  T  I  I  N  Q  D  P  D  K  I  L  T  Y

TTGCTGcCGATCACTGCCCGGTAGTCGAGGTGAACGGcGTGACCATCCAAGTCGGGAGCAGGAGGTATCCaGACGCtGTGTACtTGCACAGAATTGACCT  6800
 I  A  a  D  H  C  P  V  V  E  V  N  G  V  T  I  Q  V  G  S  R  R  Y  P  D  A  V  Y  L  H  R  I  D  L

CGGTCCTCCCATATcATTGGAGAgGTTGGACGTAGGGACAAATCTGGGGAATGCAATTGCTAAGtTGGAGGATGCCAAGGAATTGtTGGAGTCATCGGAC  6900
 G  P  P  I  s  L  E  r  L  D  V  G  T  N  L  G  N  A  I  A  K  L  E  D  A  K  E  L  L  E  S  S  D

CAGATATTGAGGAGTATGAAAGGTTTATCGAGCACTaGCATAGTcTACATCCTGATTGCAGTGTGTCTTGGAGGGTTGATAGGGATCCCCGCTTTAATAT  7000
 Q  I  L  R  S  M  K  G  L  S  S  T  s  I  V  Y  I  L  I  A  V  C  L  G  G  L  I  G  I  P  A  L  I

GTTGCTGCAGGGGGCGtTGTAAcAAAAAGGGAGAACAAGTTGGTATGTCAAGACCAGGCCTAAAGCCTGATCTTACgGGaACATCAAAATCCTATGTAAG  7100
 C  C  C  R  G  R  C  N  K  K  G  E  Q  V  G  M  S  R  P  G  L  K  P  D  L  T  G  T  S  K  S  Y  V  R

                                      t                                                           ccagctaa
                                      ▼                                                                  ▼
GTCGCTCTGATCCtCTACAACTCTTGAAACACAAATGTCCCACAAGTCTCCTCTTCGTCATCAAGCAACCACCGCACCCAGCATCAAGCCCACCTGAAAT  7200
 S  L  *
```

Fig. A2. (Contd.)

H protein

```
TATCTCCGGcTTCCCTCTGGCCGAACAATATCGGTAGTTAATTAAAACTTAGGGTGCAAGATcATCCACAATGTCACCACAaCGAGACCGgATAAATGCC  7300
                                                                       M  S  P  q  R  D  R  I  N  A

TTCTACAAAGAtAACCCCCATCCcAagGGAAGTAGGATAGTtATtAACAGAGAACAtCTTATGATTGATaGACCTTATGttTTGCTGGCTGTTCTgTTcG  7400
 F  Y  K  D  N  P  H  P  k  G  S  R  I  V  I  N  R  E  H  L  M  I  D  R  P  Y  v  L  L  A  V  L  F

TCATGTtTCTGAGCTTGATCGGGTTGCTAGCCATTGCAGGCATtAGACTtCATCGGGCAGCCATCTACACCGCAGAGATCCATAAaAGCCTCAGCACCAA  7500
V  M  f  L  S  L  I  G  L  L  A  I  A  G  I  R  L  H  R  A  A  I  Y  T  A  E  I  H  K  S  L  S  T  N

TCTAGATGTaACTAACTCAATcGAGCATCAGGTCAAgGACGTGcTGACACCACTCTTCAAaATCATcGGTGATGAAGTGGGCCTGAGGACaCCTCAGAGa  7600
 L  D  V  T  N  S  I  E  H  Q  V  K  D  V  L  T  P  L  F  K  I  I  G  D  E  V  G  L  R  T  P  Q  R

TTCACtGAccTAGTGAAATTcATCTCTGACAAgATtAAATTCCTTAAtCCgGAtAGGGAGTACGACTTcAGAGAtCTCAcTTGGTGtATcAACCCGCcAG  7700
 F  T  D  L  V  K  f  I  S  D  K  I  K  F  L  N  P  D  R  E  Y  D  f  R  D  L  t  W  C  I  N  P  p

AgAGAaTCAAATTGGATTATGATCAATACTGTGCAGAtGtGGCTGCTgAAGAgCTCATGAATGCATTGGTGAACTCAACTCTaCTGGAGaCCAgaacAAC  7800
E  R  I  K  L  D  Y  D  Q  Y  C  A  D  v  A  A  e  E  L  M  N  A  L  V (N  S  T) L  L  E  t  r  t  T

cAATCAGTTCCTAGCTGtcTCAAAgGGAAAcTGCTCAGGGCCCAcTACAATCAGAGGTCAATTcTCaAACATGTCGCTGTCCCTGTTgGACTTGTATTTA  7900
 N  Q  F  L  A  v  S  K  G (N  C  S) G  P  t  T  I  R  G  Q  F  S (N  M  S) L  S  L  L  D  L  Y  L

agTCGAGGTTAcAATGTgTCATCTATAGTcACtATGACATCCCAGGGAATGTAcGGGGGAACTTACCtAGTGGaAAAgCCTAATCTGAGCAGcAAAgGGT  8000
 s  R  G  Y (N  V  S) S  I  V  T  M  T  S  Q  G  M  Y  G  G  T  Y  I  V  e  K  P (N  L  S) S  K  g

CAGAgTTGTcACaACTGAGCATGtAcCGaGTGTTTGAAgTAgGtGTTATCAGAAATCCgGGTTTgGGGGCTCCgGTGTTCCATATGACAAACTAttTTGA  8100
S  E  L  s  q  L  S  M  y  R  V  F  E  v  g  V  I  R  N  P  G  L  G  A  P  V  F  H  M  T  N  Y  f  E

GCAACCAGTCAGTAATGATcTCAGCAACTGtATGGTgGCTTTGGGGGAGCTCAaacTCGCAGCCCTtTGTCACgGgGaAGATTcTaTCACaaTTCCCTAT  8200
 Q  P  V  S  N  D  I  S  N  C  M  V  A  L  G  E  L  k  I  A  A  L  C  H  g  e  D  s  I  T  I  P  Y

CAGGGaTCAGGGaAAGGTGTCAGCTTcCaGCTCgTCAAgCTaGGTGTCTGGAAATCcCCAACcGaCATGCAATCCTGGGTCcCctTaTCAACGGATGAtC  8300
Q  G  S  G  k  G  V  S  F  q  L  v  K  L  G  V  W  K  S  P  T  d  M  Q  S  W  V  p  I  S  T  D  D

CAGTGATAGAcAgGCTtTACCTCTCATCTCACAGaGGtGTTATCGCTGacAATCAAGCAAaATGGGCTgTCCCGACAACAcGaACAGATGACAAgTTGCG  8400
P  V  I  D  r  L  Y  L  S  S  H  R  G  V  I  A  d  N  Q  A  k  W  A  v  P  T  T  R  T  D  D  K  L  R

AATGGAGACATGCTTCCAgCAGGCGTGtAAGGGtAaAAtCCAAGCACTCTGCGAGaATCCCGaGTGgGcACCATTGAAGGAtaaCAGGATTCCtTCATAc  8500
 M  E  T  C  F  Q  Q  A  C  K  G  k  I  Q  A  L  C  E  n  P  e  w  a  P  L  K  D  n  R  I  P  S  Y

GGgGTCTTGTCTGTTgATCTGAGtCtgACAGTTGAgCtTAAAATCAAAATTGCTTCgGGaTTCGGGCCaTTGATCACACACGGTTCAGGGATGGACCTaT  8600
 G  V  L  S  V  d  L  S  I  T  V  e  I  K  I  K  I  A  S  G  F  G  P  L  I  T  H  G  S  G  M  D  L

ACAAatCCaACcACAAcAATGtGTATTGGCTGAcTATCCCgCCAATGAAGAAcCTAGCctTAGGTGTAATcAACACATTGGAgTGGATACCGAGAtTCAA  8700
Y  K  s  n  h  N  N  v  Y  W  L  t  I  P  P  M  K  N  L  A  L  G  V  I  N  T  L  E  W  I  P  R  f  K

GGTTAgtCCcaACCtCTTcAcTgTcCCAATtAAGGAAGCAGGCGaaGACTGCCATGCCCCAACATACCtacCTGCGGAGGTGGatGgTGATGTCAAACTC  8800
 V  s  P  n  I  F  t  v  P  I  K  E  A  G  e  D  C  H  A  P  T  Y  I  p  A  E  V  d  g  D  V  K  L

AGTTCCAAtCTGGTgATtCTACCtGGtCAaGAtCTCCAATATGTtTTGGCAACCTAcGATACtTCCAGggTTGAACATGCTGTgGTTTATTAcGTTTACA  8900
 S  S  n  L  V  I  L  P  G  Q  D  L  Q  Y  V  L  A  T  Y  D  T  S  R  v  E  H  A  V  V  Y  Y  V  Y

gcCCAaGCCGCTCATTTtCTTACTTTTATCCTTTTAGGTTGCCTaTAAaGGGGgTCCCcATCGAATTAcAAGTGGAATGCTTCACATGGGACcAAAAACT  9000
s  P  s  R  S  F  s  Y  F  Y  P  F  R  L  P  I  k  G  v  P  I  E  L  q  V  E  C  F  T  W  D  q  K  L
```

Fig. A2. (Contd.)

CTGGTGCCGtCACTTCTGtGTgCTTGcGGAcTCAGaaTCTGGTGgAcATATCACtCACTCTGGGATGGtgGGCATGGgaGTCAGCTGcACAGTCACcCGG 9100
W C R H F C V L a D S e S G g h I T H S G M v G M g V S C T V T R

GAaGAtGGaACCAAtcgCAGATAGGgCTGCtAGTGAAccAATCtCAtGATGTCACCCaGACATCAgGCATAcCCACTAgTGTGAAATAGACATCAGAATT 9200
E D G T N r R *

L protein

AAGAAAAACGTAGGGTCCAAGTGgTTCCCCGTTATGGACTCGCTATCTGTCAACCAGATCTTaTACCCtGAAGTTCACCTAGATAGcCCGATAGTTACCA 9300
M D S L S V N Q I L Y P E V H L D S P I V T

AtAAGATAGTAGCCATCCTGGAGTATGCTCGAGTcCCTCACGCTTACAGCCTGGAGGACCCTACACTGTGTCAGAACATCAAGCACCGCCTAAAAAACGG 9400
N K I V A I L E Y A R V P H A Y S L E D P T L C Q N I K H R L K N G

ATTTTCCAACCAAATGATTATAAACAATGTGGAAgTTGGGAATGTCATCAAGTCCAAGCTTAGGAGTTATCCGGCCCACTCTCATATTCCATAtCCAAAt 9500
F S N Q M I I N N V E v G N V I K S K L R S Y P A H S H I P Y P N

TGTAATCAGGATTTATTTAACATAGAAGACAAAGAGTCAACGAGGAAGATCCGTGAACTCCTcAAaAAgGGgAATTCGCTGTACTCcAAAGTCAGTgATA 9600
C N Q D L F N I E D K E S T R K I R E L L K K G N S L Y S K V S d

AGGTTTTCCAATGCTTaAGgGACACtAAcTCACGGCTTGGcCTAGGCTCCGAATTGAGGGAGGACAtCAAGGAGAAAGTTATTAACTTGGGAGTTTACAT 9700
K V F Q C L R D T N S R L G L G S E L R E D I K E K V I N L G V Y M

GCACAGCTCCCAgTGGTTtGAGCCcTTTCTGTTTTGGTTtACAGTCAAGACTGAGATGAGGTCAGTGATtAAATCACAAACCCATACTTGCCATAGGAGG 9800
H S S Q W F E P F L F W F T V K T E M R S V I K S Q T H T C H R R

AGACACACACCTGTATTCTTCACTGGTAGTTCAGTTGAGTTGCTAATCTCTCGTGACCTTGTTGCTATAATCAGTAAAGAGTCTCAACATGTATATTACC 9900
R H T P V F F T G S S V E L L I S R D L V A I I S K E (S Q H V Y Y
I

TgACATTTGAaCTGGTTTTGATGTATTGTGATGTCATAGAGGGgAGGTTAATGACAGAGACcGCTATGACtATTGATGcTAGgTATACAGAgCTTCTAGg 10000
L T F E L V L M Y C D V I E G R L M T E T A M T I D a R Y T e L L g

AAGAGTCAGATACATGTGGAAAcTGATAGATGGTTTCTTCCCTGCACTCGGGAATCCAACTTATCAAATTgTAGCCATGCTGGAGCCtCTTTCACTTGCT 10100
R V R Y M W K L I D G F F P A L G N P T Y Q I v A M L E P L S L A

TACCTGCAGCTGAGGGATATAACaGTAGAACTCAGaGGTGCTTTCCTTAACCACTGCTTTACTGAAATACATGATGTTCTTgACCAAAACGGGTTTTCTG 10200
Y L Q L R D I T V E L R G A F L N H C F T E I H D V L d Q N G F S

ATGAAGGTACTTATCAtGAGTTAaTTGAAGCtCTAGATTACATTTTCATAACTGATGACATACAtCTGACAGGGGAGATTTTCTCATTTTTCAGAAGTTT 10300
D E G T Y H E L I E A L D Y I F I T D D I H L T G E I F S F F R S F

CGGCCACCCCAGACTTGAAGCAGTAACGGCTGCTGAAAATGTTAGGAAATACATGAATCAGCCTAAAGTCATTGtGTATGAGACTCTGATGAAAGGTCAT 10400
G H P R L E A V T A A E N V R K Y M N Q P K V I v Y E T L M K G H

GCCATATTTTGTGGAATcATAATcAAcGGCTATCGTGACAGGCACGGAGGCAGTTGGCCACCGCTGACCCTCCCCCTGCAtGCTGCAGACACAATCCGGA 10500
A I F C G I I I N G Y R D R H G G S W) P P L T L P L H A A D T I R

ATGCTCAAGCtTCAGGTGAaGGgTTAACACATGAGCAGTGCGTTGATAACTGGAAATCTTTTGCTGGAGTGAAATTTGGCTGCTTtATGCCTCTTAGCCT 10600
N A Q A S G e G L T H E Q C V D N W K S F A G V K F G C F M P L S L

GGATAGTGATCTGACAATGTACCTAAAGGAcAAGGCACTTGCTGCTCTCCAAAGgGAATGGGATTCAGTtTACCCGAAAGAGTTCCTGCGTTACGACCCt 10700
D S D L T M Y L K D K A L A A L Q R E W D S V Y P K E F L R Y D P

CCCAAGGGAACCGGGTCACGGAGGCTTGTAGAtGTTTTCCTTAATGATTCGAGCTTTGACCCATATGATgTGATAATGTATGTTGTaAGTGGaGCTTACC 10800
P K G T G (S R R L V D V F L N D S S F D P Y D v I M Y V V S G A Y
II

Fig. A2. (Contd.)

TCCATGACCCTGAGTTCAACCTGTCTTACAGCCTGaAAGAaAAGGAGATCAAGGAAACAGGTAGACTTTTTGCTAAAATGACTTACAAAATGAGGGCATG 10900

L H D P E F N L S Y S L k E K E I K E T G R L F A K M T Y K M R A C

CCAAGTGATTGCtGAAAATCTAATCTCAAACGGGATTGGCAAATATTTTAAGGACAATGGGATGGCCAAGGATGAGCAcGATTTGACTAAGGCACTCCAC 11000

Q V I A E N L I S N G I G K Y F K D N G M A K D E h D L T K A L H

ACTCTaGCTGTCTCAGGAGTCCCcAAAGATCTCAAAGAAAGTCACAGgGGGGGCCAGTCtTaAAAACCTACTCCCGAAGCCCAGtCCACACAAgTACCA 11100

T L A V S G V P K D L K E S H R G G P V L K T Y S R S P v H T s T

GGAACGTGAGaGCAGCAAAAGgGTTTATAGgGTTCCCTCAagTAATTCGGCAGGACCAAGAcACTgATCATCCGGAGAaTATGGAAGCTTACGAgACAGT 11200

R N V R A A K g F I g F P Q v I R Q D Q D T d H P E n M E A Y E T V

III

CAGTGCATTTATCACgACTGATCTCAAGAAGTACTGCCTTAATTGGAGATATGAGACCATCAGCTTGTTTGCACAGAGGCTAAATGAGATTTACGGATTg 11300

S A F I T T D L K K Y C L N W R Y E T I S L F A Q R L N E I Y G L

A

CCCTCATTtTTcCAGTGGCTGCATAAGAGGCTTGAGACCTCTGTCCTGTATGTAAGTGACCCTCATTGCCCCCCGACCTTGACGCCCATATCCCGTTAT 11400

P S F F Q W L H K R L E T S V L Y V S D P H C P P D L D A H I P L

atAAAGTCCCCAATGAtCAAATCTTCATtAAGTACCCTATGGGAGGTATAGAAGGGTATTGTCAGAAGCTGTGGACCATCAGCACCATTCCCTATcTATA 11500

y K V P N D Q I F I K Y P M G G I E G Y C Q K L W T I S T I P Y L Y

B

CCTGGCTGCTTATGAGAGCGGAGTAAGGATTGCTTCgTTAGTGCAAGGGGACAATCAGACCATAGCcGTAACAAAaAGGGTACCCAGCACATGGCCcTAC 11600

L A A Y E S G V R I A S L V Q G D N Q T I A V T K R V P S T W P Y

C

AAcCTTAAgAAAcGGGAAGCTGCTAGAGTAACTAGAGATTACTTTGTAATTCTTAGGCAAAGGCTACATGAtATTGGCCATCACCTCAAGGCAAATGAGA 11700

N L K K r E A A R V T R D Y F V I L R Q R L H D I G H H L K A N E

CAATTGTTTCATCACATTTTTTTGTcTATTCAAAAGGAATATATTATGATGGGCTACTTGTGTCCCAATCACTCAAGAGCATCGCAAGATGTGTATTCTG 11800

T I V S S H F F V Y S K G I Y Y D G L L V S Q S L K S I A R C V F W

D

GTCAGAGACTATAGTTGATGAAACAAGGGCAGCATGCAGTAATATTGCTACAACAATGGCTAAAAGCATCGAGAGAGGTTATGACCGTTACCTTGCATAT 11900

S E T I V D E T R A A C S N I A T T M A K S I E R G Y D R Y L A Y

TCCCTGAACgTCCTAAAAGTGATaCAGCAaATTCTgATCTCTCTTGGCTTCACAATCAATTCAACCATGACCCgGGATGTAGTCATACCCCTCCTCACAA 12000

S L N v L K V I Q Q I L I S L G F T I N S T M T r D V V I P L L T

ACAACGACCTCTTAATAAGGATGGCACTGTTGCCCGCTCCTATTGGGGGGATGAATTATCTGAATATGAGCAGGCTGTTTGTCAGaAACATCGGTGATCC 12100

N N D L L I R M A L L P A P I G G M N Y L N M S R L F V R N I G D P

IV

AGTaACATCATCAATTGCTGATCTCAAGAGAATGATTCTCGCcTCACTaATGCCTGAAGAGACcCTCCATCAgGTAATGACACAaCAACCgGGGGACTCT 12200

V T S S I A D L K R M I L A S L M P E E T L H Q V M T Q Q P G D S

TCATTCCTAGACTGGGCTAGCGACCCTTACTCAGCAAATCTTGTATGTGTCCAGAGCATCACtAGACTCCTCAAgAACATAACTGCAAGGTTTGTCCTgA 12300

S F L D W A S D P Y S A N L V C V Q S I T R L L K N I T A R F V L

TCCAtAGTCCAAACCCAATGTTAAAaGGATTATTCCATGATGACAGTAAAGAAGAGGACGAGgGACTGGCgGCATTCCTCATGGACAGGCATATTATAGT 12400

I H S P N P M L K G L F H D D S K E E D E g L A A F L M D R H I I V

ACCTAGGGCAGCTCATGAAATCCTGGATCATAGTGTCACAGGgGCAAGAGAGTCTATTGCAGGCATGCTgGATACCACAAAAGGCtTGATTCGAGCCAGC 12500

P R A A H E I L D H S V T G A R E S I A G M L D T T K G L I R A S

ATGAGGAAGGGGGGtTTAACCTCTCGAGTGATAACCAGATTGTCCAATTATGACTATGAACAATTCAGAGCAGGGATGGTGCTATTgACaGGAAGAAAGA 12600

M R K G G L T S R V I T R L S N Y D Y E Q F R A G M V L L T G R K

Fig. A2. (Contd.)

GAAATGTCCTCATTGACAAAGAGTCATGTTCAGTGCAGCTGGCGAGAGCtCTAAGAAGCCATATGTGGGCGAGGCTAGCTCGAGGACGGCCTATTTACGG 12700
R N V L I D (K E S C S V Q L A R A L R S H M W A R L A R G R P I Y G
V

CCTTGAGGTCCCTGATGTACTAGAATCTATGCGAGGCCACCTTATTCGGCGTCATGAGACATGTGTCATCTGCGAGTGTGGATCAGTCAACTACGGATGG 12800
L E V P D V L E S M R G H L I R R H E T C V I C E C G S V N Y G W

TTTTTTGTCCCCTCGGGTTGCCAACTGGATGATATTGACAAGGAAACATCATCCtTGAGAGTCCCATATATTGGTTCTACCACTGATGAGAGAACAGACA 12900
F F V P S G C Q L D D I D K E T S S L R V P Y I G S T T D E R T D

TGAAGCTTGCCTTCGTAAGAGCCCCAAGTCGATCCTTGCGATCtGCTGTTAGAATAGCAACAGTGTACTCATGGGCTTACGGTGATGATGATAGCTCTTG 13000
M K L A F V R A P S R S L R S A V R I A T V Y S W A Y G D D D S S W

GAACGAAGCCTGGTTGTTGGCtAGGCAAAGGGCcAATGTGAGCCTGGAGGAGCTAAGGGTGATCACTCCCATCTCAACTTCGACTAATTTAGCgCATAGG 13100
N E A W L L A R Q R A N V S L E E L R V I T P I S T S T N L A H R

TTGAGGGATCGTAGCACTCAAGTGAAATACTCaGGTACATCCCTTGTCCGaGTGGCgAGGTATACCACAATCTCCAACGACAATCTCTCATTTGTCATAT 13200
L R D R S T Q V K Y S G T S L V R V A R Y T T I S N D N L S F V I

CAGATAAGAAGGTTGATACtAACTTTATATACCAACAAGGAATGCTtCTAGGGTTGGGtGTTTTAGAAACATTGTTTCGACTCGAGAAAGATACCGGATC 13300
S D K K V D T N F I Y Q Q G M L L G L G V L E T L F R L E K D T G S

ATCTAACACgGTATTACATCTTCACGTCGAAACAGATTGTTGCGTGATcCCgATGATAGATCATCCCAGGATACCcAGCTCcCGCAAGCTAGAGCTGAGG 13400
S N T V L H L H V E T D C C V I P M I D) H P R I P S S R K L E L R

GCAGAGCTaTGTACCAACCCATTGATATATGATAATGCACCtTTAATTGACAGAGATgCAACAAGGCTATACACCCAGAGCCATAGGAGGCACCTTGTgG 13500
A E L C T N P L I Y D N A P L I D R D a T R L Y T Q S H R R H L V

AATTTGTTACATGGTCCACACCCCAACTATATCACATTtTAGCTAAGTCCACaGCACTATCTATGATTGACCTgGTAACAAAATTTGAGAAGGACCATAT 13600
E F V T W S T P Q L Y H I L A K S T A L S M I D L V T K F E K D H M

GAATGAAATtTCAGCTCTCATAGGGGATGACGATATCAATAGTTTCATAACTGAGTTTCTgcTcATAGAGCCAAGATTATTCACTATCTACTTGGGCCAG 13700
N E I S A L I G D D D I N S F I T E F L I I E P R L F T I Y L G Q

TGTGCGGCCATCAATTGGGCATTTGATGTACATTATCATAGACCATCAGGGAAATATCAGATGGGTGAGCTGTTGTCATCGTTCCTTTCTAGAATGAGCA 13800
C A A I N W A F D V H Y H R P S G K Y Q M G E L L S S F L S R M S

AAGGAGTGTTTAAGGTGCTTGTCAATGCTCTAAGCCACCCAAAGATCTACAAGAAATTCTGGCAtTGTGGTATTATAGAGCCTATCCATGGTCCTTCACT 13900
K G V F K V L V N A L S H P K I Y K K F W H C G I I E P I H G P S L

TGATGCTCAAAACTTGCACACAACTGTGTGCAACATGgTTTACACATGCTATATGACCTACCTCGACcTGTTGTTGAATGAAGAGTTAGAAGAGTTcACA 14000
D A Q N L H T T V C N M v Y T C Y M T Y L D L L L N E E L E E F T

TTTCTcTTGTGTGAAAGcGACGAGGATGTAGTACCGGACAGATTCGACAACATCCAGGCAAAACACTTaTGTGTTCTGGCAGATTTGTACTGTCAACCAG 14100
F L L C E S D E D V V P D R F D N I Q A K H L C V L A D L Y C Q P

GGaCCTGCCCACCAATtCgaGGTCTAAGACCGGTAGAGAAATGTGCAGTTCTAACCGACCATATCAAGGCaGAGGCTAtGTTATCTCCAGCAGGATCTTC 14200
G t C P P I r G L R P V E K C A V L T D H I K A E A m L S P A G S S

GTGGAACATAAATCCAATTATTGTAGACCATTACTCATGCTCtCTGACTTATCTCCGGCGAGGATCGATCAAACAGATAAGATTGAGAGTTGATCCAGGA 14300
W N I N P I I V D H Y S C S L T Y L R R G S I K Q I R L R V D P G

TTCATTTTCGACGCCCTCGCTGAGGTAAATGTCAGTCAGCCAAAGATCgGCAGCAACAACATCTCAAATATGAGCATCAAGGcTTTCAGACCCCCACACG 14400
F I F D A L A E V N V S Q P K I g S N N I S N M S I K a F R P P H

Fig. A2. (Contd.)

```
ATGATGTTGCAAAATTGCTCAAAGATATCAAcACAAgCAAGCACAATCTTCCCATTTCaGGGGGCAATCTcGCCAATTATGAAATCCATGCTTTCCGCAG   14500
 D  D  V  A  K  L  L  K  D  I  N  T  s  K  H--N  L--P  I  S  G  G  N  L  A  N  Y  E  I  H  A (F  R  R
                                                                                              VI
AATCGGGTTGAACTCATCTGCTTGCTACAAAGCTGTTGAGATATCAACATTAATTAGGAGATGCCTTGAGCCAGGGGAgGACGGcTTgTTCTTGGGTGAG   14600
   I  G  L  N  S  S  A  C  Y  K  A  V  E  I  S  T  L  I  R  R  C  L  E  P  G  E  D  G  L  F  L  G  E

GGATCGGGTTCTATGTTGATCACTTATAAGGAGATACTTAAACTAAgCAAGTGCTTCTATAATAGTGGGGTTTCcGCCAATTCTAGATCTGGTCAAAGGG   14700
  G  S  G  S  M  L  I  T  Y  K  E  I  L  K  L  s  K  C  F  Y  N  S  G  V  S  A  N  S  R  S  G  Q  R

AATTAGCACCCTATCCCTCCGAAGTTGGCCTTGTCGAaCACAGAATGGGAGTAGGTAATATTGTCAAaGTGCTCTTTAACGGGAGGCCCGAAGTCACgTG   14800
 E  L  A  P  Y  P  S  E  V) G  L  V  E  H  R  M  G  V  G  N  I  V  K  V  L  F  N  G  R  P  E  V  T  W

GGTAGGCAGTGTAGATTGCTTCAATTtCATAGTTAGTAATATCCCTACCTCTAGTGTGGGGTTTATCCATTCAGATATAGAGACCTTgCCTgACAAAGAT   14900
   V  G  S  V  D  C  F  N  f  I  V  S  N  I  P  T  S  S  V  G  F  I  H  S  D  I  E  T  L  P  d  K  D

ACTATAGAGAAGCTAGAGGAATTGGCAGCCATCTTATCgATGGCtCTGCTCCTGGGCAAAATAGGATCAATACTGGTGATTAAGCTTATGCCTTTCAGCG   15000
  T  I  E  K  L  E  E  L  A  A  I  L  S  M  A  L  L  L  G  K  I  G  S  I  L  V  I  K  L  M  P  F  S

GGGATTTTGTTCAGGGATTTATAAGTTATGTAGGGTCTcATTATAGAGAAGTGAACCTTGTATACCCtAGATACAGCAACTTCATATCTACTGAATCTTA   15100
 G  D  F  V  Q  G  F  I  S  Y  V  G  S  h  Y  R  E  V  N  L  V  Y  P  R  Y  S  N  F  I  S  T  E  S  Y

TTTGGTTATGACAGATCTCAAGGCTAAcCGGCTAATGAATCCTGAAAAGATTAAgCAGCAGATAATTGAATCATCTGTGaGGACTTCACCTGGACTTATA   15200
  L  V  M  T  D  L  K  A  N  R  L  M  N  P  E  K  I  K  Q  Q  I  I  E  S  S  V  R  T  S  P  G  L  I

GGTCACATCCTATCCATTAAGCAACTAAGCTGCATACAAGCAATTGTGGGAGACGcAGTTAGTAGAGGTGATATCAAtCCTACTCTGAAAAAACTTACAC   15300
  G  H  I  L  S  I  K  Q  L  S  C  I  Q  A  I  V  G  D  a  V  S  R  G  D  I  N  P  T  L  K  K  L  T

CTATAGAGCAGGTGCTGATCAATTGCGGGTTGGCAATTAACGGACCTAAgCTGTGCAAAGAATTGATCCACCATGATGTTGCCTCAGGGCAAGATGGATT   15400
 P  I  E  Q  V  L  I  N  C  G  L  A  I  N  G  P  K  L  C  K  E  L  I  H  H  D  V  A  S  G  Q  D  G  L

GCTTAATTCTATaCTCATCCTCTACAGGGAGTTGGCAAGATTcAAaGACAACCaAAGAAGTCAACAAGGGATGTTcCACGCTTACCCCGTATTGGTAAGT   15500
  L  N  S  I  L  I  L  Y  R  E  L  A  R  F  K  D  N  q  R  S  Q  Q  G  M  F  H  A  Y  P  V  L  V  S

AGCAGGCAACGAGAACTtATATCTAGgATCACcCGCAAATTTTGGGGGCACATTCTTCTTTACTCCGGGAACAGAAAGTTGATAaATAAGTTtATCCAGA   15600
  S  R  Q  R  E  L  I  S  R  I  T  R  K  F  W  G  H  I  L  L  Y  S  G  N  R  K  L  I  n  K  F  I  Q

ATCTCAAGTCCGGcTATCTGATACTAGACTTACACCAGAATATCTTCGTTAAGAATCTaTCCAAGTCAGAGAAACAGATTATTATGACGGGGGGTTTGAA   15700
 N  L  K  S  G  Y  L  I  L  D  L  H  Q  N  I  F  V  K  N  L  S  K  S  E  K  Q  I  I  M  T  G  G  L  K

ACGTGAGTGGGTTTTTAAGGTAACAgTCAAGGAGACCAAaGAATGGTATAAGTTAGTCGGATACAGTGCCCTGATTAAGGACTAATTGGTTGAACTCCGG   15800
   R  E  W  V  F  K  V  T  v  K  E  T  K  E  W  Y  K  L  V  G  Y  S  A  L  I  K  D  *

                                          trailer
AACCCTAATCCtgCCCtAGGTgGTtAGGCATTATTTGCAATAtATTAAAGAAAACTTTGAAAATACGAAGTTTCTATTCCCAGCTTTGTCTGGT        15894
   B' box                                                                      A' box
```

Fig. A2. (Contd.)

Squared boxes in the nucleotide sequence mark start and stop codons, the nontranscribed gene boundary trinucleotides, and the transcription initiation and termination regions of the genes. Squared boxes in the amino acid sequences denote transmembrane regions, the presumed leader peptide (fine stipples) and the fusogenic region of the F_1 protein (coarse stipples). The arrows in the amino acid sequence of the F protein show proteolytic cleavage sites (removal of the leader peptide, generation of the mature F_1 and F_2 proteins). Ovals around a group of amino acids mark the N-glycosylation sites; sites known not to be actually glycosylated are stippled. Circles indicate the highly conserved cysteine residues in the V protein.

The rounded boxes I-VI represent regions of high conservation in the L protein of the Mononegavirales. The amino acids underlined in the box III constitute the conserved region A-D shared by RNA-dependent polymerases of the plus-strand and negative-strand RNA viruses (Poch et al. 1990).

References

Calain P, Roux L (1993) The rule of six: a basic feature for efficient replication of Sendai virus defective interfering RNA. J Virol 67: 4822–4830

Poch O, Blumberg BM, Bouqueleret L, Tordo N (1990) Sequence comparison of five polymerases (L polymerases) of unsegmented negative-strand RNA viruses: theoretical assignment of functional domains. J Gen Virol 71: 1153–1162

Subject Index

Current Topics in Microbiology and Immunology

Volumes published since 1989 (and still available)

Vol. 149: **Shen-Ong, Grace L. C.; Potter, Michael; Copeland, Neal G. (Ed.):** Mechanisms in Myeloid Tumorigenesis. Workshop at the National Cancer Institute, National Institutes of Health, Bethesda, MD, USA, March 22, 1988. 1989. 42 figs. X, 172 pp. ISBN 3-540-50968-2

Vol. 150: **Jann, Klaus; Jann, Barbara (Ed.):** Bacterial Capsules. 1989. 33 figs. XII, 176 pp. ISBN 3-540-51049-4

Vol. 151: **Jann, Klaus; Jann, Barbara (Ed.):** Bacterial Adhesins. 1990. 23 figs. XII, 192 pp. ISBN 3-540-51052-4

Vol. 152: **Bosma, Melvin J.; Phillips, Robert A.; Schuler, Walter (Ed.):** The Scid Mouse. Characterization and Potential Uses. EMBO Workshop held at the Basel Institute for Immunology, Basel, Switzerland, February 20–22, 1989. 1989. 72 figs. XII, 263 pp. ISBN 3-540-51512-7

Vol. 153: **Lambris, John D. (Ed.):** The Third Component of Complement. Chemistry and Biology. 1989. 38 figs. X, 251 pp. ISBN 3-540-51513-5

Vol. 154: **McDougall, James K. (Ed.):** Cytomegaloviruses. 1990. 58 figs. IX, 286 pp. ISBN 3-540-51514-3

Vol. 155: **Kaufmann, Stefan H. E. (Ed.):** T-Cell Paradigms in Parasitic and Bacterial Infections. 1990. 24 figs. IX, 162 pp. ISBN 3-540-51515-1

Vol. 156: **Dyrberg, Thomas (Ed.):** The Role of Viruses and the Immune System in Diabetes Mellitus. 1990. 15 figs. XI, 142 pp. ISBN 3-540-51918-1

Vol. 157: **Swanstrom, Roland; Vogt, Peter K. (Ed.):** Retroviruses. Strategies of Replication. 1990. 40 figs. XII, 260 pp. ISBN 3-540-51895-9

Vol. 158: **Muzyczka, Nicholas (Ed.):** Viral Expression Vectors. 1992. 20 figs. IX, 176 pp. ISBN 3-540-52431-2

Vol. 159: **Gray, David; Sprent, Jonathan (Ed.):** Immunological Memory. 1990. 38 figs. XII, 156 pp. ISBN 3-540-51921-1

Vol. 160: **Oldstone, Michael B. A.; Koprowski, Hilary (Eds.):** Retrovirus Infections of the Nervous System. 1990. 16 figs. XII, 176 pp. ISBN 3-540-51939-4

Vol. 161: **Racaniello, Vincent R. (Ed.):** Picornaviruses. 1990. 12 figs. X, 194 pp. ISBN 3-540-52429-0

Vol. 162: **Roy, Polly; Gorman, Barry M. (Eds.):** Bluetongue Viruses. 1990. 37 figs. X, 200 pp. ISBN 3-540-51922-X

Vol. 163: **Turner, Peter C.; Moyer, Richard W. (Eds.):** Poxviruses. 1990. 23 figs. X, 210 pp. ISBN 3-540-52430-4

Vol. 164: **Bækkeskov, Steinnun; Hansen, Bruno (Eds.):** Human Diabetes. 1990. 9 figs. X, 198 pp. ISBN 3-540-52652-8

Vol. 165: **Bothwell, Mark (Ed.):** Neuronal Growth Factors. 1991. 14 figs. IX, 173 pp. ISBN 3-540-52654-4

Vol. 166: **Potter, Michael; Melchers, Fritz (Eds.):** Mechanisms in B-Cell Neoplasia. 1990. 143 figs. XIX, 380 pp. ISBN 3-540-52886-5

Vol. 167: **Kaufmann, Stefan H. E. (Ed.):** Heat Shock Proteins and Immune Response. 1991. 18 figs. IX, 214 pp. ISBN 3-540-52857-1

Vol. 168: **Mason, William S.; Seeger, Christoph (Eds.):** Hepadnaviruses. Molecular Biology and Pathogenesis. 1991. 21 figs. X, 206 pp. ISBN 3-540-53060-6

Vol. 169: **Kolakofsky, Daniel (Ed.):** Bunyaviridae. 1991. 34 figs. X, 256 pp. ISBN 3-540-53061-4

Vol. 170: **Compans, Richard W. (Ed.):** Protein Traffic in Eukaryotic Cells. Selected Reviews. 1991. 14 figs. X, 186 pp. ISBN 3-540-53631-0

Vol. 171: **Kung, Hsing-Jien; Vogt, Peter K. (Eds.):** Retroviral Insertion and Oncogene Activation. 1991. 18 figs. X, 179 pp. ISBN 3-540-53857-7

Vol. 172: **Chesebro, Bruce W. (Ed.):** Transmissible Spongiform Encephalopathies. 1991. 48 figs. X, 288 pp. ISBN 3-540-53883-6

Vol. 173: **Pfeffer, Klaus; Heeg, Klaus; Wagner, Hermann; Riethmüller, Gert (Eds.):** Function and Specificity of Á / ‰ TCells. 1991. 41 figs. XII, 296 pp. ISBN 3-540-53781-3

Vol. 174: **Fleischer, Bernhard; Sjögren, Hans Olov (Eds.):** Superantigens. 1991. 13 figs. IX, 137 pp. ISBN 3-540-54205-1

Vol. 175: **Aktories, Klaus (Ed.):** ADP-Ribosylating Toxins. 1992. 23 figs. IX, 148 pp. ISBN 3-540-54598-0

Vol. 176: **Holland, John J. (Ed.):** Genetic Diversity of RNA Viruses. 1992. 34 figs. IX, 226 pp. ISBN 3-540-54652-9

Vol. 177: **Müller-Sieburg, Christa; Torok-Storb, Beverly; Visser, Jan; Storb, Rainer (Eds.):** Hematopoietic Stem Cells. 1992. 18 figs. XIII, 143 pp. ISBN 3-540-54531-X

Vol. 178: **Parker, Charles J. (Ed.):** Membrane Defenses Against Attack by Complement and Perforins. 1992. 26 figs. VIII, 188 pp. ISBN 3-540-54653-7

Vol. 179: **Rouse, Barry T. (Ed.):** Herpes Simplex Virus. 1992. 9 figs. X, 180 pp. ISBN 3-540-55066-6

Vol. 180: **Sansonetti, P. J. (Ed.):** Pathogenesis of Shigellosis. 1992. 15 figs. X, 143 pp. ISBN 3-540-55058-5

Vol. 181: **Russell, Stephen W.; Gordon, Siamon (Eds.):** Macrophage Biology and Activation. 1992. 42 figs. IX, 299 pp. ISBN 3-540-55293-6

Vol. 182: **Potter, Michael; Melchers, Fritz (Eds.):** Mechanisms in B-Cell Neoplasia. 1992. 188 figs. XX, 499 pp. ISBN 3-540-55658-3

Vol. 183: **Dimmock, Nigel J.:** Neutralization of Animal Viruses. 1993. 10 figs. VII, 149 pp. ISBN 3-540-56030-0

Vol. 184: **Dunon, Dominique; Mackay, Charles R.; Imhof, Beat A. (Eds.):** Adhesion in Leukocyte Homing and Differentiation. 1993. 37 figs. IX, 260 pp. ISBN 3-540-56756-9

Vol. 185: **Ramig, Robert F. (Ed.):** Rotaviruses. 1994. 37 figs. X, 380 pp. ISBN 3-540-56761-5

Vol. 186: **zur Hausen, Harald (Ed.):** Human Pathogenic Papillomaviruses. 1994. 37 figs. XIII, 274 pp. ISBN 3-540-57193-0

Vol. 187: **Rupprecht, Charles E.; Dietzschold, Bernhard; Koprowski, Hilary (Eds.):** Lyssaviruses. 1994. 50 figs. IX, 352 pp. ISBN 3-540-57194-9

Vol. 188: **Letvin, Norman L.; Desrosiers, Ronald C. (Eds.):** Simian Immunodeficiency Virus. 1994. 37 figs. X, 240 pp. ISBN 3-540-57274-0

Vol. 189: **Oldstone, Michael B. A. (Ed.):** Cytotoxic T-Lymphocytes in Human Viral and Malaria Infections. 1994. 37 figs. IX, 210 pp. ISBN 3-540-57259-7

Vol. 190: **Koprowski, Hilary; Lipkin, W. Ian (Eds.):** Borna Disease. 1995. 36 figs. IX, 196 pp. ISBN 3-540-57388-7

MIX
Papier aus verantwortungsvollen Quellen
Paper from responsible sources
FSC® C105338

If you have any concerns about our products, you can contact us on
ProductSafety@springernature.com

In case Publisher is established outside the EU, the EU authorized representative is:
Springer Nature Customer Service Center GmbH
Europaplatz 3, 69115 Heidelberg, Germany

Printed by Libri Plureos GmbH
in Hamburg, Germany